RIPPLES

RAYMOND HEINS

PUBLISHER'S INFORMATION

EBook Bakery Books

ISBN 978-1-953080-47-9
Author contact: rayheins@yahoo.com

DEDICATION

To Jack Galvin – friend and mentor, who led me (and many others) through the personal essay process. He rightly judged that everyone has stories to tell. By the time we were finished with the journey, we felt that we knew each other as friends, not just classmates. To bring our stories to light, Jack provided empathy, tools, and encouragement, as well as examples from his own experience.

Jack served his community for as long as he lived, beginning with Rogers High School, then Salve Regina University. No one who worked with him will ever forget his guidance, kindness, and inspiration.

RIP

1939 – 2023

INTRODUCTION

Newport's Salve Regina University has an ongoing studies program for senior citizens, called the "Circle of Scholars". As part of the program, author and professor Jack Galvin offered a course in personal essays, and the essays in *Ripples* were the result of this extraordinary academic offering from an extraordinary man.

Course sessions were eight weeks long. During each of those weeks, Galvin offered guidance from his lengthy career as an English teacher (and department head) at Newport's Rogers High School. He also provided the topic for each new essay. Crucially, the second half of each week's session was reserved for a student to read his or her essay from the previous week. Topics included "The Fight in You", "A Meaningful Song", "A Struggle in Your Life", "A Place", "A Value You Learned the Hard Way", "A Significant Person" – Galvin's list seemed endless.

Writing a four page essay each week was laborious but after listening to eight soul-searching essays, the six or so classmates felt closer to each other than life-long friends. To the question: Why publish your own scribblings? Ray responds:

For Vanity – I think some of the essays are well done. (It is your job, dear reader, to identify which ones!) I felt the insights into my own psyche were hard-earned.

For Self Introduction – This is who I am, and have been, for the friends I know and those I've never met.

For Family – This is who I was, before age and geography separated us.

For Fellow "Inmates" in Assisted Living – This is who I used to be, before the transformations of age and health.

In any case, I hope that you, the reader, will remember that there are hidden stories in each of us and every one of your acquaintances - just as there are in you, yourself. Wouldn't it be nice if everyone would dredge up a few of their defining memories, and commit them to a written form?

Raymond Heins, November, 2023

Contents

CHILDHOOD

1

A Fifty Cent Theft

I received my first instruction on race relations when I was six years old. It was probably overdue, as we lived in the 1940's South. It was prompted by an incident at the corner grocery store, where I had been sent to make a minor purchase for my mother. I was sent with a fifty cent piece to cover the shopping. A very nice colored man was the only other customer in this store. He smiled, and asked

"What brings you to this store, young man?"

"We need a loaf of bread," I responded, happy to be noticed.

"And did you bring any money?" he asked.

"Yes sir!" - I showed him the nice, substantial half dollar I had brought.

"Oh, my - that's pretty! Can I show you a trick? Let me see that half-dollar."

I happily complied, and he demonstrated how the coin could magically disappear. Then it re-appeared, and he plopped it into my shirt pocket. I was very impressed with his trick.

"'Bye now" he said; "I have to go to work."

"Goodbye, sir", I responded. Then I asked Mr. Blunkhall, the grocer, for the bread - but I could not find the money - I had lost it. Crestfallen, I returned home empty-handed.

When I got home, mother soon learned that I had made a new friend, while losing track of the money. She seemed most disturbed that I had been chatting with a black man, and that

I had called him "Sir".

"Don't you know he stole your money?" She seemed very angry. "Listen to me: Don't you ever call a black man Sir!" she commanded. "If you have to talk to one of *Them*, you call him by his first name. Better yet, don't even talk to Them. Just stay away!"

This scolding was hard to take in. All my life, up to that point, respect for adults had always been demanded. I was not at all sure I would be able to follow her primary command. What if I didn't know his first name? But I decided I could avoid Them. Looking back, probably she was warning me to stay away from strangers - but if so, it was clear that the most dangerous of unknowns would have dark skin.

My mother did not hate Them. She readily acknowledged some talents. When it came to hard physical work, such as digging out a stump, a white man simply could not measure up to a colored one. She used at least two mispronunciations of "Negro", one more respectful than the other, but would never have insulted a Negro by calling him or her "black". Their skin was always some shade of brown, never black. Nevertheless, They were the kind of people who would take money from a small boy, and were never to be trusted.

As the next several years went by, I had virtually no contact with Them. I would see Them, sometimes, stocking shelves, cleaning, or laboring outside. They had special facilities, like water fountains and rest rooms, just for "COLORED ONLY". They also had their own restaurants, schools, and, I suppose, stores. On the bus, they got to ride in the back, including the one extra-wide seat at the very back. That, too, was reserved by a "COLORED" sign.

My first eight years of education were in a Catholic school, populated exclusively by whites: Sisters, Lay teachers, and students. This never received the slightest notice from anyone - it was the natural order of things, and we would no more expect colored people in our school than a green sky and blue grass at recess.

And then came May, 1954. A unanimous decision by the U.S. Supreme Court declared that "separate but equal" wasn't good enough. The education received was not, in fact, equal, and schools would have to become "integrated".

My parents were sure the country was going to hell in a hurry, but this decision probably would have no immediate effect. Only it did. The Bishop of Nashville declared that all Catholic schools in the diocese would integrate immediately - that fall - September 1954. The schools were open to all comers. What a shock! We wondered if there would be room for any white students.

In fall I registered with the only Catholic boys' high school in Nashville, Father Ryan. Surprisingly, only a few black boys registered to attend. Or maybe it wasn't so surprising. Who would willingly plunge into an environment where everyone was hostile to you? Surely that thought occurred to Them, but plunge they did - at least a few.

So the next four years began my secondary education in race relations, as well as math, English, and science. There was little socializing, but I gradually realized from classroom contact that They were just as variable as whites. Some were smart, some not so much. Some were sloppy, some neat. They all talked funny ("Matthew" came out "Matchoo"), but I decided not all of Them would steal money from a child.

Social interaction was limited, partly because our world was not completely in our control. All our athletic rivals were local public schools, and colored players were not eligible to participate by league rules. I do not remember any black students attending school dances; if They were there, there was an unstated pact of mutual invisibility in effect. A group of the black students formed a vocal group called the Thunderbirds, and I do remember attending at least two performances. They were very good at Doo-Wop Rock 'n' Roll, which was, of course, very popular throughout white America in the 50's.

Slowly, we became less sensitive to Their presence. Some of my more adventurous friends (white, of course) led me into some closer encounters. Once we went to a R&B concert in a converted supermarket building. Nearly all other attendees were black, but everyone was there for the music, not trouble. On another occasion, I was one of the representatives from my high school to attend a half-day math contest held at Pearl High School - which was exclusively black. There, I got a little taste of life as a black student in a white school, except that everyone was very friendly.

Every Southerner would agree that black folks were especially good at two things, at least: music and barbecue. Both my family and my friends had a favorite barbecue source: Brown's BBQ. It was located in Their section of town, and the owners were indeed brown. Usually we would buy a pound or so of the product and depart. But on one occasion, a small group of friends and I craved a late night barbecue sandwich, so we stopped in at the restaurant section. Obeying custom (non-Catholic school custom), we stood at the lunch counter, waited for our sandwiches, then departed. We did not feel welcome to sit down, just as blacks were not welcome to sit in a white restaurant. It was an interesting experience.

A year after graduating from high school, I received an appointment to the Naval Academy, reporting for Plebe Summer July 7, 1959. Among many shocks, I learned definitively that one of Them could be in charge! Thus began my undergraduate education in race relations, although it would not take four years to finish. After two months of intense physical and mental training as a member of a fifteen man squad of plebes, I became one of the 30 or so fourth classmen (a.k.a. plebes) in a 100 man Company of all grades. At dinner that first night, I found that my twelve-man table included two first class midshipmen (seniors), two second class, four third class, and four plebes.

Dinner for plebes is always stressful. You must sit "braced up" on the front four inches of your chair, and respond to harassing questions, even while ensuring the upperclassmen are well served at all times. Nourishment for the plebes is absolutely the last thing on the agenda. Most questions require regurgitation of factoids learned from "Reef Points", which contains distilled wisdom from and about naval service. But one of my two second classmen was Willie Zeal Byrd - a very dark-skinned and very hostile native of Alabama. He probed intensely, just hoping he could detect the slightest hint of disrespect from me, the plebe from Tennessee. I had no difficulty calling him "Sir", several times in one sentence, if necessary.

Looking back, I think of that meal as being my commencement exercise from the school of race. At the Naval Academy, and later in the Navy fleet, and later still in civilian life, I never again had the slightest difficulty working with and respecting officers, petty officers, ordinary seamen, or civilians on the basis of race. "They" became "they", and I never told any of them about my fifty cent theft.

2

Danielsville, Georgia

In childhood, I used to visit a magic place, pretty much every year. Not everyone would think it magic, or even habitable. There was no electricity or plumbing. We used an outhouse, which was equipped with old newspapers and old Sears Roebuck catalogs – dual purpose amenities. The house was so quiet, with no radio, no refrigerator, not even any fans, that people could talk in their soft, quiet voices, from one room to another.

This magic kingdom was set in rural northeast Georgia, and it was home to my mother's family. She was one of seven children born to James Marion Rice and Jincy Caroline (Callie) Hooper. Mother (Blanche) was clearly the most adventurous, as she had moved, as a young woman, 250 miles north (a challenging distance in the 1920's) to Nashville to study nursing at Vanderbilt University. The rest continued to live in and around Danielsville Georgia – a dusty crossroads whose downtown consisted entirely of a single building – a combination general store and post office. The store was located on a paved road, giving it a certain elite status. The farmhouses occupied by Mother's sisters - Aunt Flora and Aunt Mamie - were reached by hot, dry, dusty roads bordered by red clay shoulders wherever the roadway dipped below field level.

My most memorable visit took place in 1952 – the year I was twelve. Cousin John Frank Varner was attending Peabody

Teacher's College in Nashville, and I was allowed to ride down to Georgia with him in his Studebaker ("The front end looks just like the back end – you can't tell if it's coming or going!") a week or ten days before the rest of my family would arrive for their visit.

The crops were in, so there was no heavy labor for me to do – or maybe they simply spared me. My most taxing chore was to draw and carry water from the circular stone well, with spindle, rope, and bucket operated by a crank at the end of the spindle. There was plenty of time and space for walks, through woods or along dusty roads, exploring the creeks, catching little frogs. Once I lost my way, and the tight, panicky feeling was unpleasant, but I decided if I just walked along the dirt road, I would eventually wind up somewhere I could ask for directions – and instead I found Aunt Flora's house where I was staying.

Another adventure was getting to ride the mule. Clearly this was not a customary use of those patient, strong, utilitarian animals. The family did not even own a saddle, so I rode bareback. It turns out that a mule's spine stands up about four inches above any meaty portion of his rib cage, so it was a most uncomfortable ride! I was even happier to disembark than to climb aboard.

In memory, that idyllic life was marked for me by its quiet. Those Georgia farmers spoke gently and slowly. They had time to consider every utterance, and there was never any need to speak above another voice, or any competing sounds. Time moved slowly. Every day was pretty much like every other, except maybe for Saturday – bath day. It didn't occur to me to wonder where anyone else bathed, but for me it was on the back porch, in a steel laundry tub, warmed with water from the wood-burning cook stove. Strangely, I cannot remember a Sunday. Mother and we were Catholics, during an era when you surely would perish if you ever set foot inside a Protestant church. The Georgia Rice clan all were Methodist, I think, but they were so considerate the question of attending church simply never came up. They probably knew about our mortal danger, should we accompany them, and didn't even suggest it.

One evening Cousin John Frank was going into town, and I was invited along. We went, not into Danielsville, but into the bigger town of Commerce, Georgia. The streets were paved and lighted. There was even a movie theater. I must have acclimated to farm life very quickly, because this little town was dazzling to my eyes. John Frank bought me a box of Butterfinger nuggets – a strikingly generous present, which I soon greedily devoured. Those surpassingly delicious, sweet, sticky nuggets have never been topped by any treat, any time, anywhere.

All those people are gone now. Both Aunts, their husbands, most of the cousins, and even John Frank – who was killed at age 30 while taking down a tree. Their way of life is gone, as well. Electricity and plumbing came soon enough, as the Tennessee Authority (TVA) extended its reach. I'm sure pavement, road traffic, radio, television, refrigerators, and probably the sound of airplanes have ended the reign of silence. But I still think, occasionally, of that magic place, and am thankful that I had the privilege of visiting.

3

The Change Dish

I remember the Change Dish as clearly as though I saw it this morning, even though it was really more than 70 years ago. It rested on top of the "sideboard", as we called it. (I never believed that was a correct name – it seemed more of a chest or cabinet, to me.) Anyway, this cabinet had a top surface maybe five feet wide, two feet deep. Mother always kept a white cloth runner on it. At the right end, nearest the doorway from the hall into the dining room, was the Change Dish. A few ration tokens continued to live in that dish, even though the war had ended long ago – a year or two. More importantly, loose change was kept there. Pennies, a few nickels, a quarter or two, and maybe three or four dimes. I especially loved dimes. So small, so thin, and yet capable of purchasing two whole candy bars. The best thing about the Change Dish was that it never ran out. You could spend from it day after day, and yet there would still be money there.

And I did have the opportunity to spend. Magnusson's grocery was located at the intersection of 25th and Capers Avenues – only about six doors from our house, although it was across both streets – diagonally from "our" corner of 25th and Capers. But Mother would send me to the store routinely. As a six-year old, I was proud to be entrusted with this task. After all, when the leaves were out, you couldn't even see the store from our house.

Mother needed me to go to the store because of Judy. That little interloper, three years old now, took up an awful lot of Mother's attention. Usually that was a bad thing, but when I got to go to the store, I didn't mind.

"Raymond, I need you to go to the store, and get us three-quarters of a pound of ground round." That was a very frequent assignment. It didn't occur to me that ¾ of a pound wasn't much meat for a family of six. Cooked with cabbage and tomato sauce, it made a family favorite – "Dago Cabbage", she called it. I had no idea that name would have been offensive to some people, and I don't think Mother knew it either. The ethnic origins of people simply didn't occur to me. No one I knew seemed any different from anyone else, except maybe in their distance from farm life. The Sprys next door seemed embarrassingly close to farmers. They spoke of "chicken aigs", and "w'are p'lars" (wire pliers) and sounded, I thought, like hillbillies.

At the store, Mr. Magnusson knew me, and made sure I got what I needed. Usually I must have returned home in less than 15 minutes after setting out – although I never timed a trip.

Eventually, I conceived the idea of using that Change Dish for my own purposes. Before setting out for the store, I was always given approximately the right amount of money, but an extra dime from the Change Dish offered great reward. I had the motive (I loved chocolate), the means (I could reach the Change Dish) and the opportunity (my trip to the store) – and so I committed the crime. My, those Hersey bars were good!

It was the next day when disaster struck: "Oh, no!" said Mother, with a note of alarm in her voice. "There's money missing from the Change Dish!" I was the only person in the room, so I assume she was talking to me. But wise beyond my years, I kept my mouth shut.

"I can't imagine what happened, but there is a dime missing." She mused. "I know my son didn't take it – he would never do a thing like that." Stricken, I continued to keep my silence, but hot shame was flooding over me. I was sure she was going to pursue the crime to conviction and punishment, but she stopped, still staring at that Change Dish. Maybe she wasn't sure how much had been in there! I began to hope, at least. I had not realized how important every nickel and every dime was until now.

Miraculously, the inquisition was not resumed, but I was determined never again to breach a trust. At first it was the Change Dish, but I soon found it extended to a wider range of experiences. I can't say there were never lapses, but my outlook on the world shifted a bit. For the rest of my life, Honesty was more than a virtue; it was a goal, a practice, almost a compulsion. I am incapable of flattery, for example, even if I sense it is a lack in my interpersonal arsenal. A semi-friend of mine once called me "the most honest man I know" – and I'm not sure it was a compliment. Today I always report an undercharge, or too much change, to the cashier. (That doesn't mean I hesitate to point out an overcharge, of course.) The closest I can come to dishonesty is omission – I still know how to keep my mouth shut, on occasion.

The importance of honesty has been reinforced, of course, first in school, then in Scouting, then in military service, finally in parenting. There were elaborate "honor systems" at both Vanderbilt University and the Naval Academy. Plebe Year harassment at the Academy also emphasized honor and worthiness of trust. But my Mother's homespun psychology on a six year old child was as powerful as the rest of them put together.

TEEN YEARS & COLLEGE

4

A Not-So-Typical English Teacher

"STAND UP, BOY! WHAT'S YOUR NAME?"

"Raymond Heins, Father"

"WERE YOU TALKING IN MY CLASS?"

"No, Father..."

"SHUT UP! I'LL DO THE TALKING IN HERE. YOU TALK WHEN YOU'RE CALLED UPON; GET IT?"

"Yes, Father..."

This conversation was accompanied, once I came to my feet, with several emphatic pokes to the chest. Since he had seized my shirt, the emphasis motion flopped my head back and forth. The entire classroom was staring, and utterly silent.

It was my first English class, Freshman year, at Father Ryan High School, and I had now met Father Morgan. And I had

NOT been talking, but that mattered not at all. Even at the time, I realized that Father Morgan probably put on this little show at the first meeting with all Freshman English classes. I had heard of him while still in grade school. "He's really tough", I was assured by classmates who had older brothers at Ryan.

We had nuns for teachers during most of our careers in Grades 1 to 8. They could be tough, too, and had no trouble cracking knuckles when they felt it had been earned. But the fearsome reputation of the priests at Father Ryan added a sense of dread – we were going to a different level of education in more ways than one.

Father Ryan was an all-boys school. With a student body of a little over 300, there were about 75 graduates each year. Father Ryan was the only Catholic boys' high school, fed primarily by the half-dozen Catholic Grade Schools in the Nashville area. There was plenty of hands-on discipline when needed, but that first day was my only experience with it.

Father Morgan was a stocky man, probably 5'10" tall, weighing in at maybe 180 pounds. Like all the priest teachers, he wore a black cassock with "Roman collar". His hair, cut short and receding, was not gray; he must have been around 40, although that age seemed quite advanced to a high school freshman. He had the look of an ex-athlete, maybe a boxer. His nose seemed to have been re-arranged. His voice carried a tough guy edge, and no detectable accent (to my Tennessee ears, at any rate). He demanded rapt attention from his students, and tolerated no back-talk or side talk.

As soon as he had our attention on that first day, he announced that we knew nothing whatsoever about English grammar, so he was going to begin again, teaching us everything we should have learned in Grade School. I soon realized he was absolutely correct. We began with subject, verb, object; nouns, pronouns, adjectives. None of these were new concepts, really, but they seemed a lot clearer when presented in an organized, related fashion – in an attentive classroom.

I remembered struggling with the concept of the "verb to be" back in sixth grade. I simply couldn't understand how a word would, perhaps, become a verb. Is it or isn't it? What changes it? What is it while it's not a verb? I found it hard to formulate questions about my confusion, and the teacher certainly couldn't clarify it for me. Finally, in Father Morgan's class, this and many other parts of English mechanics were revealed.

As the year progressed, I realized that I appreciated the controlled classroom he maintained. Some of the other teachers, despite frequent bluster and occasional chastisement, failed miserably in controlling the class. Spitballs flew, messages and written games were passed, every time the teacher faced the blackboard. Books fell from desks. Whispered jokes prevailed. Knowledge, it seemed, didn't thrive in certain classrooms.

As I progressed through high school, that was the only class I took from Father Morgan. Either he specialized in Freshman English, or he taught courses I wasn't exposed to. My English teachers in subsequent years picked up where he left off – we had an understanding of the mechanics, so they could expand on literature, composition, and the like. Those courses seemed at once more enjoyable and easier than many others – Mathematics, Latin, Chemistry, Physics, Religion, Spanish for example. History was, of course, simply impossible – useless and dull, lost in the dead distant past prior to my birth. So English class was usually a relief, a pleasure.

Following high school, I took another year of "English, the subject" at Vanderbilt, then one at the Naval Academy. After that year, English was folded into a composite (English, History and Government – affectionately known among midshipmen as "Bull"). Writing skills, however, continued to be useful – highly significant in grade achievement in many courses. (Not all, of course; one roommate could misspell his own name, but understood scientific concepts so well, he was able to achieve a dual degree in our four year stay by the Severn River.)

Upon graduation, I entered the Navy world. Many different skills were used, but writing ability was always helpful in such areas as drafting a convincing letter, writing subordinate's performance reports, and even drafting naval messages with omitted articles, pronouns, adjectives, and adverbs. Writing skills hammered into my skull by Father Morgan contributed to promotions, to success in postgraduate school, to success in my shore tours.

After retirement from active duty, writing skills continued to be a big part of my duties as a computer systems development manager, and later as an appraiser. Convincing, coherent arguments and descriptions were always useful, and always a source of satisfaction when I read them over. It's been a long time since Father Morgan introduced himself to me, but I still remember him very well, indeed!

5

FRAULEIN

Composer: Lawton Williams
Singer: Bobby Helms

Far across deep blue waters, lives an old German's daughter
By the banks of the old river Rhine
Where I loved her and left her, but I can't forget her
I miss my pretty Fraulein.

IT'S HARD-CORE COUNTRY - no denying that. I first heard it while it was a big hit, in 1957. It was one of the most popular selections on the juke box at Nashville's Scorchy's Diner, on 25th Avenue, just a few blocks from my high school. Scorchy's was a "real" diner - a converted trolley car - with lots of chrome and plastic and a boxy addition at the entrance. Scorchy's wife, Ruth, was the face, providing table service while Scorchy ran the kitchen.

"Fraulein" was a song of lament for lost love - voiced by a GI who had met and loved a girl in Germany, probably during occupation duty after World War II. The lyrics urge the girl to "look up towards Heaven, each night when the stars start to shine." And it promises that, "by the same stars above you, I swear that I love you."

The occupation had ended just five years earlier, so this wasn't ancient history. In fact, American troops remain in Germany to this day.

Scorchy's provided mugs of ice cold beer for twenty five cents, and the juke box provided six songs for a quarter, one for a nickel. Even then, these prices were excellent! But the best thing of all was that there were no questions asked.

My friends and I were all seventeen, give or take, and of course we had no business swilling beer - especially on lunch break from school. Eventually there was a crackdown on exactly our type of behavior, and suspensions resulted. But not in our year. So, we ordered a beer and a sandwich, listened to the juke box, discussed girls, and hustled back for our 1:00 o'clock class. We had our youthful dreams of personal romance, and our dreams of life after high school. Would we go off to college? Would some of us join the army and see the world? Maybe we would meet one of those pretty Frauleins.

Well, years did pass, most of us did not become alcoholics or criminals, and we did return, occasionally, to the diner. Scorchy and Ruth broke up, and we were sure Scorchy was at fault. Ruth kept the diner going, even moving it to a new location (this time, an actual building). The prices went up, a little, but they still were reasonable. The juke box continued to feature "Fraulein", and our thoughts evolved, as well. Now I pictured a real German girl, seen on one of my navy port visits, who was as idealized and unattainable as the musical one.

The group aged, and dispersed. David joined the Air Force for a few years. John and Mike worked as surveyors. I spent a year in Engineering school, then departed for the U.S. Naval Academy. Frank spent some time as a miscreant, then became a policeman. Big Ed developed acromegaly, a pituitary disorder which results in enlarged features, but eventually became a Marshall.

One Christmas Eve, around 1962, we boys assembled in Ruth's while light snow fell outside. An aging pair of alcoholic, nearly derelict twins, John and Paul, were regulars at the time. They usually preferred a cheap wine known as "Sweet Mary", but in a pinch would accept beer donations. John had slipped and fallen on the sidewalk outside. He was moaning piteously: "I think my laig's broke!" Ruth knew him to be a man who would do anything to cadge a drink, but felt his performance deserved a reward. First she provided him a glass of succor, then we followed. Still, John kept up his complaints - when he wasn't drinking his beer. Finally, possibly just to get him out of the diner, Ruth called for a medical pickup (this was before the days of "911"). Weren't we surprised when it turned out he really had broken his laig!

Some of the Guys at Ruth's Diner - Age 33 (Ed, Mike, Dave, John)

The last time I visited the Diner (photo above) was in 1973. "Fraulein" was still on the juke box, but now it cost a quarter for just one song. My group never discussed what it meant to us individually, but certainly each of us linked it to the group, and to our vanishing youth.

Probably more than anyone else in my circle, those bygone days continue to have a hold on me. I moved away permanently after my father died (1976) and my mother moved to live nearer family. Now once in five years or so I will come for a visit; those who remained in the city continued their lives, while I began a new one. That old life, though, lives on in my unconscious mind. When I hear the inimitable voices of George Jones, Johnny Cash, or Bobby Bare, I think of Nashville and my life there. And when I hear "Fraulein" (yes, it's still available), I think of Ruth's Diner.

6

YOU CANNOT GO HOME AGAIN

NOTE: Several times in my writing assignments, I addressed the question: "Can You Go Home Again?" As it turns out, there are many possible answers. This one focuses on an actual place I once called Home. The title answers that question!

I GREW UP AT 2513 Capers Avenue, an address I learned almost as soon as I could talk. Our house was a "bungalow", according to my mother. It was wood clapboard painted white when I was really young, later gray with white trim. There was a full-width concrete porch, with four brick-based columns supporting the porch roof - also full width. There were four bedrooms - two downstairs, two in the finished attic, and there was one bath, occasionally supplemented by "the bushes" in moments of dire need.

All floors were pine, plentifully equipped with large, vicious splinters imperfectly suppressed with coats of wax. Walls were plaster downstairs, homasote (fiber-board) upstairs. My brother once demonstrated that he could put his fist through the wall inside his closet. The hole was there for the rest of the structure's life.

There was a basement, half dirt, half concrete, centered on the coal furnace. That furnace could and did heat the entire house, using what I now see was architectural genius. The furnace provided generous, even overpowering, heat directly above it, through a four-by-four steel grate in the center hall. That hall opened to both downstairs bedrooms, the living room, dining room, and the solo bathroom. It also sent heat up the stairway to the attic bedrooms, if the stair door was open.

The house was purchased shortly after my parents married, in 1927, for $2,500. The value remained basically unchanged for the next thirty years. But through all those years, my mother regretted the fact that it was not a brick house.

I remember that house mostly in vignettes:

Christmas: Santa's gifts, unwrapped to distinguish them, arrived on the brick hearth in front of the living room fireplace. Our stockings (the largest sock we each owned) hung from the mantle above that hearth. The Christmas tree was also in the living room, of course, as were sofa, chairs, and the console model radio/record player. Those Santa gifts usually featured a doll and some form of tea set for Judy (three years younger than I), a cap gun for me and my brother (the more realistic, the better), and usually one other special gift I had chosen from one of the toy catalogs during the run-up to the big day. Once it was a twelve dollar chemistry set - I was barely able to hope Santa would be that generous.

Another year, the best possible present, an Cocker Spaniel puppy! I never expected nor received a gift to top that one.

Dining Room: An oval-shaped, heavy cherry table, with a couple of leaves always in place, was surrounded by six dining chairs. These were assigned to family members in strict order. The smallest (Judy) nearest to mother's chair at the foot of the table; Daddy at the head of the table, Gordon and Jimmy on the north side, my mother and I on the south side. The kitchen was behind me, a tea cart in the corner behind mother, a side board behind Gordon and Jimmy, and a beautiful cherry corner cupboard behind Daddy's chair. One summer supper in 1954, after Gordon had graduated from college, moved to Milwaukee, and married, my mother read us the letter with wonderful news: Gordon and Ellen were expecting! A baby! She took a photo of the family reacting to the joyous news; that dining room image remains in my mind to this day - even if the photo is lost.

Gordon's Room: In his college years, Gordon (ten years older - I idolized him) had his bedroom upstairs, at the east end of the house. He had a big, roll top desk for his studies. In the window was a large, whole-house fan exhausting hot air from the attic. Depending on what windows and doors were open, it could provide some relief to the entire house, or more to the upstairs rooms. My other brother, Jimmy, and I shared the second upstairs bedroom. I remember Gordon studying very late at night, long after the rest of us had gone to bed.

Winters: We would have two or three good deep snows each year. On rare occasions, we (father and sons) would go sledding on Reservoir Hill. The steeply inclined roadway was useless to autos, and sledders would come from all around. Those thrilling flights on our Western Flyer sleds made wintertime a joy. At home, I remember the parents shoveling snow (both Mother and Daddy on occasion - although he did it much more often.) Strangely, I don't remember ever being drafted into service.

The Back Yard: Although our lot was only 50 by 150 feet, it seemed large enough to me. Over the years, it hosted our (my) dog, and several different gardens. During and after the war, we had a chicken coop, which provided us with both meat and eggs for many years. There were trees for climbing, room for a rifle (BB) range, and room for throwing - footballs, baseballs, and a steel sphere we thought of as a shot put. (In hindsight, it may have been an ancient cannon ball!). In short, with a little imagination, there was room for adventure.

The Basement: Daddy had a heavy wooden work bench, with what seemed a vast array of tools. We were never reprimanded for playing with them, so long as we didn't bleed. This was also where my mother would bring the children at certain times when she felt the southwest sky looked threatening. She was sure that a tornado was coming, and that it would come from that direction. In the basement, near the outside door, we were prepared for anything including a house collapse. Happily, we never had the chance to find out, but it was exciting to contemplate.

Progress came to Capers Avenue, eventually. The city of Nashville exercised its power of eminent domain, commanding that every property in the neighborhood be conveyed to Vanderbilt University. Compensation was worked out with all the property owners, and Mother finally would get to live in her brick house. By this time, I was in school at Annapolis, so I didn't get to witness the process, or hear all the details. By the time I graduated, in 1963, my parents were still living in the house but ownership had passed to Vanderbilt. There was a family gathering to mark my graduation and entry into the Navy. It was also a farewell to my childhood home. Maintenance had already ceased, and every property in the neighborhood was looking dilapidated. Not too long after our farewell, everyone had moved out of the neighborhood, and houses were demolished.

Today, our entire property is one small element of a practice soccer field. Many memories live on, and occasionally we former children will drive to the spot we believe to be in front of where the house was, but there is no "there" there. Even the street name is gone; "Capers Avenue" is now "Children's Way." With the scattering of family, passing of the parents, and total disappearance of the old home place, I am forced to admit that, sometimes, "You cannot go home again."

Somewhere near the center of this photograph, all those memories were created. Today, nothing – not the alley behind our house, nor the driveway cutout of the curb, nor even the topographic slope of the street, yields the slightest evidence of our home or the homes of any neighbors. No one can return to this home!

7

The Quest for Riches

NOTE: This responded to an assignment to write about a quest or search; later in the course we were assigned to write about our attitude toward money. This one could have counted for both!

When did I first become aware of money? I couldn't have been more than six years old. I was dispatched by my mother, on occasion, to the corner grocery for meal makings - frequently "three quarters of a pound of ground round." While on these errands, I became aware that a nickel could purchase a candy bar; money could be pretty useful!

Before I was ten, I was "allowed" to help out at my father's upholstering business. sweeping floors, "picking tow" (pulling apart previously used cushioning material to allow its re-use - not sure what "tow" was), sometimes helping to remove worn covers from chairs. My favorite task was making cloth-covered buttons which would be used to secure tufts. My primary task, probably, was to be out from underfoot at home during school vacation days. But the best part of the job came on Saturday, when I received my own small, brown envelope with cash inside. It wasn't very much money - less than a dollar when I was small - but it was MINE.

My mother took me to the bank to open a passbook savings account. I began to dream about what I could buy when the cash built up.

As I grew, I did find many ways to spend money, whether it came from my labors or from my parents. There was a difference, though. I didn't feel I needed to justify spending my own money, whereas family money (for shoes, clothes, or school supplies) really was spent by my mother.

My purchases were more interesting, I thought. I had a bicycle, handed down from my older brother. It seemed to need tires, tubes, and peddles on a regular basis. I would wangle a trip to Western Auto where I could find all these supplies, and also BB's - needed, of course, for my hand-me-down Daisy BB gun.

Still, I had aspirations: I would have loved to own a new Schwinn bike, but I didn't have enough money, and my parents wouldn't even consider it. Likewise for the BB gun; mine was probably made sometime around the civil war, I felt, and there were much better air rifles on the market. Some you could pump up, for a higher-velocity shot. Now THAT was progress - but there was no money for it. Just as there was no money to replace our 1939 Plymouth, which was so outdated, I dreaded rainy days because my mother would drive the Plymouth to pick us up from school.

As a kid in elementary school, there was simply no way I could see to earn money. That didn't stop me from wanting it, but the dream was deferred until I graduated from eighth grade, and applied for a paper route. That was a sure way, I felt, to begin my road to riches. This job persisted throughout my freshman year in high school. I awoke, thanks to my trusty hand-me-down Big Ben, at 4:30 A.M. A bike ride of a little over a mile got me to the newspaper sub-station, where a dozen or so routes were served. I would finish my route in time to go to school, but then I had to hustle home for the afternoon route. Back to the substation, around the route, home by 5:00 or so. Fridays I had to collect from every home on the route, and make appropriate entries in a special ledger. The job took a lot of hours, but by the end of the month, I had earned approximately $40.00. Some of that money went into the bank account, a down payment on college, and my mother's idea. But the rest was "walking-around money" which I treasured.

As my sophomore year approached, I resigned from my paper route, because I really, really wanted to play football, and I was getting big enough to do it. But I then "inherited" a job at the drug store soda fountain from my older brother James. Luckily for me, he had been reliable and industrious, so the pharmacy owners were happy to hire me when he left. This job paid just $0.50 per hour, but was weekends only, so it didn't interfere with football. I was pleased to learn it had certain unanticipated perks: on Sunday mornings, two girls would turn up from their nearby (Presbyterian) church for sodas and chat. Eventually I escorted one of them - the pretty one - to a sorority dance; but I wished I had connected with the vivacious one instead. It turned out the pretty one didn't want her parents to know too much about me. Was it because I went to the Catholic high school? Or was it that I had no money and no prospects?

After about a year as a soda jerk, I became a supermarket sack boy instead. More money ($0.65 per hour), and more hours - but still flexible enough to allow for football. This job lasted through one year at Vanderbilt University.

Throughout these employment experiences, I always wished for more money. What I had was enough to allow me to walk around with something in my pocket, and I did buy certain high-fashion clothes: a red nylon jacket like the one James Dean wore in "Rebel Without a Cause", and blue suede half-Wellington boots, ideally worn with orange woolen socks. My football teammates and I paraded downtown between morning and afternoon practices the summer before my senior year, and I felt on top of the world - but still craved a car of my own. My friends were just as cash-strapped as I was, but one of the boys had access to a 1951 Chevrolet which had no brakes but could get us from place to place.

As my senior year of high school progressed, I became more aware of the need for college. Both brothers had attended Vanderbilt University, located near enough to walk from our house to school. Gordon had worked for a year between high school and college, so his cash needs were minimal. James had a plum job in a local hospital's pharmacy, where he both earned money and had study time. I had my grocery store job, but felt the money pinch more acutely than ever in my life. For starters, many of the Vanderbilt students were well off; fraternity membership and collegiate partying did not come cheaply. I needed to work whenever I could to pay for books, supplies, and tuition. As a result, I felt no affiliation at all with the university, never attending even one football game, or one basketball game.

At the end of one semester, my money was gone. My parents could help, but not very much so I went to the Bursar's office to ask for a loan. After reviewing my record and financial statement, I was amazed to hear the school would waive all fees - so long as I maintained a B average!

This arrangement temporarily alleviated my youthful quest for cash, but replaced it with a concern about grades. Holding that "B" average was far from a given. I was pursuing an engineering degree (both brothers had become electrical engineers, and were recruited on-campus into very well-paying jobs). Required courses in calculus and chemistry each seemed determined to undermine any aspirations I had for a high grade point average.

There was a better way, I was sure. Before enrolling at Vanderbilt, I had applied for an NROTC scholarship. The Naval Reserve Officer Training Corps would pay for tuition and books, with the sole requirement that I achieve passing grades - no "B average" requirement. Sadly, I was not chosen for that scholarship. Nevertheless, I had enrolled as a non-scholarship participant in NROTC. I received uniforms, attended an interesting naval history class, and looked forward to a summer cruise on a navy ship. At this point, midway through freshman year, I re-applied for the NROTC scholarship, and also began pursuing an appointment to the naval academy. Still, I maintained my employment at the supermarket. It was my sole source of cash, but it did maintain my isolation from campus life. I had never attended a Vanderbilt social event, and had no school friends. What social life I had was still focused on high school friends, or on co-workers at the store.

Several months after applying, I was notified that I had won not one, but two lotteries; both naval academy appointment and NROTC scholarship were mine for the taking. I was very impressed by the color brochure from the naval academy, showing marching midshipmen in full dress uniforms. Not only was tuition covered, but midshipmen were paid! I was sold.

I had never even seen an ocean, so I cannot say I joined the navy to go to sea. Nor was I signing up to defend my country. I was aware that this would be a full time occupation, in which I would be on an identical footing with every other student. As it turned out, absolutely everything was provided - even underwear, shoes, winter coats, rain gear, haircuts - especially haircuts! Nothing was retained - we shipped our suitcases, containing all the clothes we had traveled in, back home on day two of our residence in Annapolis. For the first time since age six, I had no need to even think about money. Until that first payday, anyway; when I learned we would receive precisely five dollars per month. Yes we were all on the same budget, but good grief! I earned more on my paper route! Once again, I began to feel the need for MORE...

8

Collision Circle

I am standing on the open bridge of a destroyer. Another destroyer churns the sea a thousand yards away. The other ship, brilliant sunshine and gusting wind do nothing to distract from the voices on the padded earphones.

> BRIDGE COMBAT; SONAR DETECTION BEARING ZERO NINER ZERO, RANGE ONE THOUSAND YARDS. EVALUATED POSITIVE SUBMARINE. DESIGNATE WOLF ALPHA.
>
> BRIDGE AYE; ALL ENGINES AHEAD FULL! COME RIGHT TO ZERO NINER ZERO.
>
> ALL STATIONS BRIDGE; PREPARE FOR HEDGEHOG ATTACK.

Clarity and calm are everything on these sound powered phone exchanges. No discussion, no questions, no panic. Everyone must become a part of a large purposeful machine. Each speaker has to enunciate clearly and speak loudly enough to ensure the transducer in his mouthpiece generates an electrical signal in the earphones of all network stations. As the ship approaches firing position, commands resume:

> ALL STATIONS BRIDGE, STAND BY WEAPONS; MARK, MARK MARK

(Weapon Launch is visually confirmed by Bridge personnel)

> RIGHT FIFTEEN DEGREES RUDDER.

(Verbal command from me as the Conning Officer to the Helmsman)

During the summer of 1962, when I was a Midshipman First Class (i.e. a rising senior at the U.S. Naval Academy), my summer training was carried out on USS *Bache* (DD 470) at Guantanamo Bay Cuba. The ship served during WW II as an escort vessel for convoys in the Atlantic. She was a Fletcher class destroyer, fast and lean; 376 feet long, 2,050 tons, 60,000 shaft horsepower, top speed "in excess of" 30 knots. Originally fitted with five guns (5"/38 caliber), she had been modified to concentrate on anti-submarine warfare when I knew her. In exchange for three of those guns, she had picked up two Hedgehogs mounts (rocket-thrown depth bombs) and an additional anti-sub weapon with the mysterious name of "Weapon ALFA".

Most Navy men were familiar with Hedghogs in the 1960's. It was a semi-fixed platform with 24 "spigots". Mounted on these spigots were rocket-launched depth bombs designed to explode when they made contact with a submarine target. This was a step forward from depth charges so familiar from WW II movies; even though each depth charge was much larger than an individual hedgehog depth bomb, the depth charge weapon system was much less lethal to submarines. One major disadvantage was the fact that a depth charge explosion provided excellent hiding space for a surviving submarine. The ship's sonar simply could not penetrate the disturbed water, so the sub could change course, speed and depth to evade further prosecution.

An attack using hedgehog weapons produced a 100 foot circular pattern of splashes, 250 yards ahead of the attacking ship. The projectiles plunged through the water at 23 feet per second, exploding only when making contact with the submarine. With a sub at 200 feet depth, the explosion would take place within ten seconds of entering the water.

The beautiful thing about this weapon was this: my ship, engaging the submarine, could maintain sonar contact for extended periods of time. Immediately after firing the hedgehog, a sharp turn to starboard would keep the submarine in the sonar detection envelope. Two ships could prosecute the same submarine continuously, tag team style; the contact, directly ahead when we launched weapons, could be maintained on my port side by a starboard turn immediately after firing. He remained in the effective search zone, so long as I didn't overrun him or let him drift into my baffles (which began about 20 degrees abaft my beam, and extended throughout the stern zone). After I turned away, our cooperating partner could make his attack run, while I cleared to the right, then shifted rudder, to re-open the range without letting the target drift into the baffles where the propeller disturbance would hide him. The result was a pin wheeling circle, 1,000 yards in diameter, with the submarine marking the center. The two ships, keeping up a speed of approximately 20 knots, were continuously turning toward each other with a clearance of just a thousand yards. (With a combined speed of 40 knots, it would take well under one minute to arrange a horrendous collision!) And I had the privilege of driving our ship.

Wearing a set of sound powered phones, I was in contact with the Combat Information Center for a surface plot, and with Sonar for the underwater plot. In my mind, I could keep the overall picture up to the second. I gave voice commands to the helmsman to control our ship's path. I was aided in this by a display on the bridge, showing the relative bearing and range to the submarine contact. Much of the thrill was in constantly avoiding a collision with a fast moving ship! It may not surprise anyone to learn that my ship's captain sat in his bridge chair, essentially at my elbow, and supervised me like a hawk!

This exercise also required a cooperative submarine, of course. Each time we "fired" our weapons, we (Sonar) notified the submarine we had fired, using the "Underwater Telephone" code named Gertrude. the submarine would eject an air bubble, which (hopefully) came to the surface in the same location as the dye patch released by our exercise weapon.

I didn't really think - then - about the enormity of the trust being exhibited by that captain, allowing a 22 year old midshipman to drive his ship in such a hazardous maneuver. I had previously shown I could be relied upon to speak on the sound-powered phones loudly, clearly, and with the correct terms. I had no doubts about my ability to understand and execute the attack, but for this stranger to agree with me now seems remarkable.

It is notable to realize those waters were to be the scene of the Cuban face-off between the U.S. Navy and the Soviet Navy, just three months later, in the Cuban Missile Blockade. The training we were receiving was not employed, Thank God, but I like to think that our ability and readiness to use that training was one of the reasons we were never called upon to do so.

Whatever happened to Weapon Alfa, you may ask. Alas, it was not as fearsome as its mysterious name promised. Only a few ships were ever equipped with the system, and no projectiles were ever fired in anger. The mount fired a single depth charge, eight feet long, weighing 525 pounds with a 250 pound explosive charge. It had a longer range than the hedgehog, 800 yards compared to 180 yards, and it was more easily re-loaded, but it shared the depth charge limitations of water disturbance from the explosion. Ultimately it was replaced by the ASROC (Anti Submarine Rocket) which launched a homing torpedo out to a range of six miles.

The ship versus submarine dance proceeded for years – probably does to this day – but it's hard for me to imagine a more engaging version of the contest than the one I played at age 22.

The USS *Bache* DD 470

The Hedgehog, a forward-throwing
anti-submarine weapon.

9

"KEELER VOSHA"

KIELER WOCHE, OR KIEL WEEK

EARLY ON IN my naval career, my first ship USS *BIGELOW* (DD 942) visited Kiel, Germany to participate in "Kiel Week" 1964, celebrating the Kiel Canal, linking the North Sea at Brunsbüttel to the Baltic Sea at Kiel-Holtenau. At the time, this visit to a German city was festive, not in any way connected to World War II. After all, the war had ended almost 20 years earlier, when I was just five years old. Still, I imagined that ordinary Germans would be timid or deferential to American military forces. And so it was that I was mildly shocked by the hearty "Guten Morgen" called out by the burly, middle-aged line handler at the North Sea entrance to the Canal. His voice suggested he was in charge, equal to anyone on our warship.

USS Bigelow transiting Kiel Canal.

The Kiel Canal (German: Nord-Ostsee-Kanal, literally "North-[to]-East [Baltic] Sea canal", known as the Kaiser-Wilhelm-Kanal when it opened in 1870, is a 60 mile shortcut from the North Sea to the Baltic. Sailing around Denmark's Jutland peninsula instead would add some 250 miles of cold, stormy travel to any vessel sailing between the Atlantic and the Baltic Sea. Despite its age, the canal is still very active.

Our visit was part of a regional annual event, during which many navies visited. Most of the vessels appeared to be in the Destroyer category. Nests of at least four abreast were created along the piers of Kiel's harbor. BIGELOW, being both American and of recent vintage (built 1957), enjoyed a pierside berth, with three other USN ships outboard.

A nest of four destroyers attending Kiel Week

Sailors from different nations mingled freely during the daily Liberty hours, and many hat swaps were executed. (Most other nations outfitted their sailors with caps incorporating a ribbon with the ship's name and nationality. US caps seemed less interesting, as every sailor on every U.S. ship wore the same unlabeled "Dixie Cup".)

One interesting practice was shipboard tours, conducted by

members of the hosting ship who spoke (for us) English. In fact, English was a common language for the ships of all nationalities. This may have been a necessary accommodation to the weakness of American education, since international NATO operations were dominated by the numbers of United States ships. Most European naval officers spoke English; virtually no Americans spoke French, German, Dutch, Norwegian, Turkish, etc. I'm pretty sure nothing has changed in that department.

In any event, a small group and I visited a German Destroyer, similar in size to BIGELOW, and even newer. We saw the bridge and selected topside spaces. Most interesting, I felt, was the space which corresponded to our CIC - Combat Information Center. Many display panels seemed more advanced than our own. There was one curtained-off area which our tour guide said was reserved for a forthcoming magnetic submarine detection system. This really impressed our little party – there was no such detection system in the U.S. Navy. Evidently it impressed the tour guide's supervisor, as well. He seemed to be giving a tongue lashing to the poor junior officer who was our guide. Evidently that information was considered sensitive, and not authorized for foreign eyes and ears. Perhaps they remembered a German version of a WW II slogan in the U.S.: "Loose lips sink ships!"

A collapsed concrete submarine pen

The next day, a bus was arranged for interested parties (including me) to see a submarine memorial at the Kiel harborfront. On the way, the bus stopped to let us see a very impressive sight – submarine pens from WW II. These included concrete roofs extending out from shore far enough to hide and protect several subs. The roof was some ten feet thick (probably 3 meters), and was impervious to the bombs of many raids during the war. At war's end, we were told, British sappers blasted the roofs so that they collapsed into the water's edge, a sloping barrier occupying the space where once floated many potent war ships.

Next stop was the memorial itself. It told the story of the U-boat campaign. There were roughly 50 German subs when the war started, and a similar number when the war ended. But during the period 1939-1945, 739 U-Boats were lost, along with 30,002 crew members. As an American seaman, I had distinctly mixed feelings, but I did feel a deep respect for those young sailors who went to sea facing such odds. The accompanying image shows the memorial as it existed before WW II, but conforms to my memory as of 1964.

Memorial to the
German submarine force

That evening, there was a reception for visiting officers at an Officers' Club (the Marinefleiger Klub). 1964 membership included German pilots of NATO Maritime Patrol Aircraft; also attending were several retired German naval officers. At least one of them proudly stated that he flew against allied shipping on the Murmansk run. I had read of the devilish conditions faced by crews of the ships delivering war materiel to the Russians. They faced lethally frigid waters, mines, submarines, and plenty of aviators such as our host, now 20 years older.

This same retired officer inadvertently revealed a bit of lingering hostility, I thought. The band had played a piece of music, mutually enjoyed by the American guests and the German hosts. In a brief discussion of the piece, I offered my expectation that "they are bound to play that again, before the night is out." His response seemed angry: "Who says they must?" he barked. "Are they bound by ropes?" Clearly his translation of "bound" was a little different from mine!

Later in the week, I had the opportunity to visit the seaside

– surprisingly to me, there was a sandy beach on this cold, gray coastline. A small number of hardy visitors were relaxing on blankets, but I cannot remember seeing anyone actually in the water. On the way to and from this beach visit, we passed through downtown Kiel. It was not a large city, but I was impressed by all the green space. Then I realized this green space must have been provided courtesy of Allied bombers!

After a week of fellowship and German beer, we departed for additional port visits to Stockholm and Amsterdam, but we didn't re-visit the Kiel Canal. The map below is included to illustrate our cruise, but is not identical. It illustrates the location of the Kiel Canal, Stockholm and Amsterdam. This Northern Europe cruise could have been improved by a stop in the UK, but I can't complain. Just six months earlier, we completed a Mediterranean cruise in which we saw Spain (Barcelona, Palma de Mallorca), Italy (Genoa, Rome and Naples), Greece (Athens), the Middle East (Beirut, Persian Gulf), Pakistan (Karachi), Ethiopia (Masawa), and Gibraltar just before our December transit to home port. I had joined the navy to see the world, and got a slam bang start, thanks to my first ship, *USS BIGELOW.*

VIETNAM

10

A MEMORABLE SONG

Lyrics: "DETROIT CITY"
– Mel Tillis, 1963
Recorded by Bobby Bare

I want to go home, I want to go home,
Oh Lord, I want to go home,
Last night I went to sleep in Detroit city,
And I dreamed about those cotton fields and home...

"Oh Lord, I want to go home!" It wasn't the first time I'd heard the song, but it struck a particular resonance when I heard it in Saigon.

It was summer, 1968. I'd been married since January, in Vietnam since March. My duties involved 30-to-45 day deployments from Fleet Command HQ, on the Saigon River, to coastal patrol zones. I was the only American on a Vietnamese navy gunboat – HQ-611 – as the advisor to the commanding officer.

I found the Vietnamese language very difficult. Despite a five-week, 200 hour immersion course with native speakers, I could do little more than say hello, and tell a taxi driver to drive straight, turn left, or turn right.

Between deployments, I had daily contact with other advisors at the Fleet Command headquarters. On patrol, however, I felt quite alone. No wonder Captain Quynh once observed to me that "You seem very melancholy."

There were many bars in Saigon, but they were sketchy, expensive, and not very appealing to me as a newly-wed. The officer's club, on the rooftop deck of the Hotel Rex, was a lot more like it. The crowd there frequently numbered close to a hundred; security was such that everyone was able (and required) to remove their personal weapons at the entry (still on the roof level, but outside the bar). On occasion, they had musical groups – usually Korean, for reasons unknown to me – who performed American songs, reportedly learned completely phonetically. They had very convincing twangs, and could have been readily accepted in Nashville, but apparently understood nothing of what they were singing.

The most popular song they did was "Detroit City", with its "I want to go home" refrain. When we sang along with it, which was almost always, we really meant it!

Having grown up in Nashville, country music had seeped into my blood, even during the years when country wasn't cool. The accents and the phrases were similar to those I heard from family, friends, school mates.

The "want to go home" theme certainly captured a longing shared by every American in Vietnam. Even the volunteers (yes, there were many of us) were frequently lonely, and fully aware of the anti-war sentiment among the population back home. Sometimes it seemed the chaos back in the states was equal to that in-country. While in Saigon between patrols, I read of the assassinations, first of Martin Luther King, then of Bobby Kennedy, in the Armed Forces newspaper "Stars and Stripes". It seemed surreal; "How can this be?" we thought, and we asked each other. (No one I knew favored Kennedy; indeed we feared he would immediately abandon the Vietnamese we had come to admire; but murder was far beyond the pale; who would be next???)

Early on in my year in Vietnam, I purchased a portable radio/

cassette recorder. This I took with me on patrols. Sometimes at night, as the gunboat patrolled offshore, I brought the radio up on deck, and listened to music programs broadcast over the AFRTS – Armed Forces Radio and Television Network. This is where I really felt alone, and the homely refrain of "Detroit City" expressed my feelings very well indeed. (I don't remember hearing Robin Williams' character from "Good Morning, Vietnam", but I did listen to his radio station.)

Now the same music transports me back to those melancholy days when I was just 28. I've long since made it back home, but somewhere along the way, lost my youth. From here, it doesn't seem quite so bad back there in Detroit City. There were moments of adventure, and unparalleled opportunity to see very foreign places. The people were generally friendly, and happy to have us there defending them. I walked alone (wearing a six-gun with cowboy-style gun belt) in villages far from Saigon, and children gathered as to a magnet. On many occasions, I relaxed with other advisors, over dinner, or beers, or over coffee. While in Saigon, I had virtually no duties, so many hours were spent playing board games in the Fleet Command building.

Still, there was that constant desire to "go home". Every serviceman was given a "DEROS" upon arrival – his Date of Estimated Return from Overseas. The tour was twelve months – no more, no less (unless you were a marine – 13 months – or something unfortunate happened). We kept that in mind always. When we heard the song, and sang along with it, I think everyone felt less alone – united, ironically, by the desire to be somewhere else

Relaxing at Poulo Damas Island

Author visiting a friendly village in Viet Nam

Vietnamese crew on parade
receiving medals from the commanding officer

Vietnamese gunboat, HQ 611

11

AIR RAID IN THE TONKIN GULF

"I hate the mid watch. Relieve at 15 minutes before midnight, off at 03:45. With breakfast at 7:00, 'Turn to' at 8:00, there's not much sleep to be had. Oh well, at least I'm better off than the Sailors – with 40 men in a berthing compartment, they can't even choose to sleep in till 8:00."

THESE MUSINGS COURSED through my head as I stood the mid-watch, as "Evaluator" in CIC – the Combat Information Center. It was December, 1966. PARKS was assigned to North SAR station in Tonkin Gulf, maybe 20 miles southeast of Haiphong, North Vietnam. Somewhere nearby was our loose companion, USS *Richmond K. Turner*, DLG 20. TURNER was one of those incredible misfits spawned by the deep thinkers of the Pentagon – a good sized warship, at least twice our size, with missiles fore and aft. Great for air defense, but... "Oh My God, we forgot the guns!!!" So this marvelous warship needed the protection afforded by a 25 year old relic of WW II, and her four five-inch guns. Just in case a goddam sampan decided to sink her.

And the North Vietnamese did have armed sampans, but also torpedo boats, a well-hidden air force, and even, if you could believe it, an adventurous biplane which could attack at very low altitude. So said Naval Intelligence – more great thinkers?

"Mr. Heins!" – it was RD1 Stone, the senior Radarman. A really squared-away sailor. Late 20's, with maybe ten years in the Navy. Intelligent, industrious, serious. "On the ball".

"Yes, Stone? I tried to shake off the lethargy accumulated over the last hour and a half.

"Jolly Green checking in." Jolly Green was the call sign of the USAF patrol plane patrolling the Gulf tonight. We had very little contact with these planes. They were Spooks in a way, laden with electronic intercept equipment, I thought. They had the capacity to carry plenty, that's for sure. If they stayed aloft. A big, prop-driven, triple-tail Super Constellation, like those flown by TWA and Pan Am in the 1940's and 50's.

Our purpose in being here – more properly the purpose of North SAR, as we were really there mainly to protect the more capable ship – was Search and Rescue in the event of an over-water bailout following an air strike on Hanoi or Haiphong. CIC was jammed with electronic equipment, including surface search and air search radar consoles, maybe twelve different radio stations, and three big Plexiglas status boards – one for a short range surface plot, one for a longer range air plot, and tabular plots for contacts, call signs, radio freqs, and events such as air strikes expected later in the day.

American advisors off duty in Saigon

Two separate spaces contained ASW and ESM equipment – Antisubmarine Warfare and Electronic Signals Monitoring. The Gulf was not deep, but we believed it possible for Russian and Chinese subs to operate. It was hard to believe they would bother with us, but we still had to watch. Electronic Support Measures (ESM) intercepts were theoretically the best way to obtain early warning of attacks. You can see a radar emitter, for example, on your ESM long before there is enough of an echo for the transmitting platform to see you. Unfortunately, there was nothing automatic about ESM, and quality of the operator was everything. We had one man, RD1 Delatte, who was really good at it.

CIC, and the supporting spaces, were equipped with a lot of air conditioning. Nowhere near enough when the spaces were jammed with people at Battle Stations, but now, with no sun and a normal watch, it was delightfully cool.

Twenty feet forward of CIC, on the bridge, Lieutenant (Junior Grade) Riley Whitman sniffed the heavy tropical air. There was essentially nothing to see, inboard or out. No contacts, except for TURNER a mile off to Starboard, and a sprinkling of stars bright enough to penetrate the haze. The Quartermaster maintained the ship's position on the provided chart table, illuminated by the red-filtered desk lamp. The helmsman steered the course ordered by Whitman, and another sailor stood by the Engine Order Telegraph ready to issue orders for the engineering spaces in the unlikely event of a change in speed. Port and Starboard lookouts on the bridge wings gazed ever and only outward – seeing nothing except the TURNER. The ship was going nowhere, just maintaining comfortable steerageway. Our orders were to remain within twelve miles of center of station, so six knots was plenty of speed.

My Evaluator station consisted of a comfortable chair, a small desk area, and a communications bonanza. I had selector switches which could join me to any radio circuit, or any of the sound-powered phone circuits on the ship. Of course, at this time of day, the only one likely to be in use was the 1JV – with talkers on the bridge, in CIC, and in Main (Engine) Control. The lookouts and Radarmen worked closely together on that circuit, to correlate any visible ships with the radar blip seen on radar. Range, bearing, course and speed were plotted on one of the status boards, along with the projected Closest Point of Approach – CPA.

Coffee and cigarettes fueled most of the people on watch. "Butt kits" were strategically mounted everywhere, including on the bridge and in CIC. The bridge was even provided with electric cigarette lighters, mounted on the bulkhead and operated with a little lever. (Using an open flame would ruin night vision for long minutes.)

"Intermittent surface contact, bearing 350 true, range 40 miles."

"That's OK – probably a fishing vessel – watch and report" I responded. "Talker, let the bridge know."

Fishing vessels weren't really that common, because they never knew if they would be judged hostiles – but usually they were safe if they avoided coming anywhere near the warships.

"Any updates?" I asked after a couple of minutes

"No sir – contact is in a fade." "Wait – there he is again, range 35 miles"

Based on the change in location, either the contact was flaky, or this fishing vessel was moving out – at more than 100 knots! Well, with just two glimpses, I'll bet on flaky.

At 03:05, the operator reported a solid track; "Contact bearing 340 degrees true, course 170, speed 150; evaluate low flier!"

Nothing was visible on our air search radar, despite the close range. I picked up the radio handset:

"TURNER, THIS IS PARKS; CONTACT BEARING 335 YOUR POSITION, RANGE

THREE FIVE MILES, EVALUATE LOW FLIER. CAN YOU CONFIRM? OVER"

Turner had nothing on radar. A figment of imagination? Flight of geese? Or that damned biplane?

"Stone, can you see any IFF?" "Identification, Friend or Foe" was a system which is supposed to help you distinguish between hostile and friendly contacts. The radar platform triggers the tracked target to send this identifying signal, and it is displayed on the radar console at the detecting site.

"Negative IFF, sir."

This is preposterous! I don't want to wake up the entire ship for a flight of geese!

At 3:09, we still had the contact, which was becoming more clear with every sweep of the radar. Still heading in our general direction, still at 150 knots (plus or minus), still no IFF. "BRIDGE, COMBAT - TELL THE CAPTAIN!"

And seconds later, I made the call: "SOUND GENERAL QUARTERS!"

The terrifying klaxon, unchanged from all those WWII Navy movies, sounded in every space of the ship. After what seemed forever, but was really about 15 seconds, the Boatswain's Mate of the Watch barked into the 1MC:

"GENERAL QUARTERS, GENERAL QUARTERS, THIS IS NOT A DRILL. SET CONDITION 1AA"

Then back to the klaxon.

Within seconds voices were joining into the sound-powered phone net, reporting spaces manned. One minute thirty seconds after the alarm, the ship was manned and ready for whatever might come. Fire control was locked on the target. Guns were loaded and synched to the fire control signals. The Captain, in his bathrobe and slippers, was in CIC, gazing intently into one of the radar consoles. Then…

"RANGE TO TARGET 8 MILES - IFF - SIR I HAVE AN IFF RETURN!"

"GUN CONTROL HOLD FIRE - TARGET MAY BE FRIENDLY"

This seemed a sensible precaution at this juncture.

"Sir, Jolly Green reports transiting our location – says his IFF may be a little weak", reported RD1 Stone.

That bastard! Now he tells us. Too low to track on air search, too far out for surface search radar, I had no idea he would be overflying us on his way south.

"SECURE FROM GENERAL QUARTERS"

So now it's nearly 3:30. I'm off watch at 3:45. Everyone's up already…

"Stone: who's my relief? "

"Mr. Jordan, sir – Weapons Officer"

Great – I dial down to his GQ station – "Hey Wes – you're up next – why don't you come on up and relieve me now?"

"Screw you, Heins – I'll see you in 15 minutes!"

12

SUNK

DAWN HAD COME and gone. I was a US Navy Lieutenant, the "American Advisor", and the only American on the vessel. I remained asleep in my upper bunk, or half asleep, lulled by the thrumming engines and steady roll of the Vietnamese gunboat, as it plodded northwest at twelve knots just off the coast of Phu Quoc island. It was Sunday morning, and nothing could possibly go wrong with the simple transit to patrol station.

BANG! Followed by more BANGS! The ship's bell rang with each BANG. The 40mm gun, I was pretty sure. It was almost directly overhead, on the foredeck. "I never knew it made so much noise", I thought, as I leapt out of bed, heart racing, and dressed. GI green tee-shirt, matching shorts (modified from fatigue pants in Saigon), flip-flops, six gun, and I was out the door.

"What the hell is with that bell?" I shouted. No one paid me the slightest attention. Evidently the bell was being sounded by the shock of the gunfire. But the entire vessel was shuddering with the BANG's, so that some effort was needed to climb the ladder to the pilot house, overlooking the foredeck. The pilot house was also directly below the 01 level and its smaller gun mounts and the infernal ship's bell.

Lots of activity. Sailors running in all directions. More BANG's, more bells. One of the sailors, the mess cook, had climbed to the 01 level and manned the .50 cal machine gun.

He was firing bursts at no target, so far as I could see. The 40mm gun remained on centerline, un-manned.

"We have struck a reef!" reported the Captain, Dai-Uy Quynn.

So... no gun, just pounding of hull on rock. And the resultant shock was indeed sounding the bell with every impact. The Captain was trying to work the vessel off the reef, alternately ordering engines full astern, and full ahead.

I returned to my cabin, grabbed the PRC-25 radio, returned to the main deck outside the pilot house, and set up the antenna clear of nearby superstructure.

"SPOTTED MOOSE THIS IS NOVEMBER JULIET", I called. After two repetitions, I had a response from the Coastal Zone headquarters watch stander.

> "NOVEMBER JULIET HARD AGROUND, GRID COORDINATES CHARLIE MIKE 075-165. REQUEST FORWARD TO FLEET COMMAND HQ. REQUEST ASSISTANCE."

"ROGER OUT" was the only response.

By now, reported Petty Officer Nguyen, sea water was beginning to collect in the aft berthing compartment. The Captain ordered Nguyen and a Damage Control Party to immediately begin de-watering.

Now two US Navy Swift Boats were arriving from other nearby patrol sectors. The Lieutenant (Junior Grade) Skipper of the first to arrive offered to take off crewmen, if the gunboat was in danger of sinking.

"No! said Captain Quynn. "We will save the boat."

I hustled below, gathered clothing, camera, and correspondence into my sea bag, but left the packed gear out of sight insofar as possible. The radio and pistol I kept at hand. Edible stores I left in the cabin.

Suddenly the lights went out; up to this point, I had given

no thought to the fact that the engine had remained operational, driving the generator. Now the deafening silence emphasized the wash of sea water on the hull. And there was now a pronounced settling of the stern. Maybe we will need that lift, after all, I thought. At least the goddam bell had stopped ringing, as the vessel bounced less.

The engineer officer reported to the captain, in Vietnamese, that the water had risen to a level that continued operation of the generator was dangerous.

The damage control team had rigged portable (gasoline powered) fire pumps on the deck, near the entrance to the crew's quarters. Large intake hoses were taking suction from the berthing compartment, each pumping 150 gallons per minute from the vessel, over the side.

"The pumps are holding steady", reported the captain for my information. "When tide increases, we hope to be free of reef."

And then what, I wondered, as I stared at the navigation chart. Do we settle on the bottom in 20 feet of water, rather than on a reef three feet below the surface? The reef upon which the ship had impaled itself was not uncharted, exactly. It showed up as having a minimum of six feet of water above it, but obviously that was outdated.

The mess cook had ceased his firing of the .50 caliber gun after the arrival of the Swift Boats. With Viet Cong-controlled jungle less than two thousand yards away, we might need those bullets yet.

Back in the berthing compartment, the pumps kept clogging up, on floating pillows, mattresses, clothing, towels – that berthing compartment had not been exactly "ship shape". Soon it was obvious, the ship was not going to be re-floated anytime soon. After two hours of attempted de-watering, the damage control party had to admit defeat. The pumps were shut down, and the compartment evacuated.

"SWIFT 71, REQUEST YOU COME ALONGSIDE",

I radioed. All but certain the ship would be completely lost within hours, I off-loaded my gear, but remained aboard to await the outcome.

One or more Swifts could remain in the vicinity indefinitely, and if the gunboat decided to break in two and sink, I was confident I and the crew could be picked up. Every man knew where his life jacket was, and some wore one continuously.

EPILOG: The ship sat firmly on the reef for seven long days and nights, until the arrival of a US Navy salvage ship. Divers patched the holes, pumps dewatered the compartments, and, after two days of work, the salvage ship towed the gunboat back to Saigon – a journey of almost 1,500 miles. The highlight of the trip, for me, was the rainfall as we were towed up the Saigon river; it was my first shower in fifteen days. Upon arrival at Fleet Command headquarters, I was reunited with my seabag, permitting me to change clothes for the first time in fourteen days. Arriving at headquarters with a two-week beard, another advisor commented: "You look like Charlton Heston returning from an Indian raid!"

13

The Volunteer

It may be little noted in other states, but Tennessee is proud to be known as the Volunteer State. Thus the University of Tennessee fields teams known as "The Volunteers". And long before I was ten years old, I was a huge fan of the Double A baseball team, the Nashville Vols. So I was fully aware, and I knew that the name came from the historic tendency of Tennesseans to volunteer for wars. Revolutionary, Civil, World - you name it, Tennessee was standing in line to enlist. This awareness played no small role, I am sure, in my decision to volunteer for in-country service in Vietnam.

I had been in naval service for several years before the U.S. became involved in southeast Asia. As a midshipman I was on a destroyer in Cuban waters during 1962, before the missile crisis and Blockade. Another destroyer, where I served as Communications Officer, participated in the "Quarantine" of the Dominican Republic in 1965. I was on that same ship when I heard the news of the Tonkin Gulf Incident in 1964. That incident was the effective trigger for US involvement in Vietnam. Two years later, I was joining my next destroyer as the Engineering Department Head, as it prepared for deployment to the western Pacific and Vietnam. At that point, a call went out from the Bureau of Naval Personnel for volunteers to serve in-country in Vietnam. The Navy had many billets for junior officers, on rivers, coasts, and both ashore and afloat in advisory positions.

Why, you may ask, would I even consider volunteering to go? 26-year old me might answer "This is My War. It's already two years old, and I wonder how long it will continue. Wars don't come along all that often, and it would be a shame to miss it altogether. After all, I'm from the Volunteer state. Furthermore, I've watched plenty of World War II movies. Many of them take place in the wide Pacific. Those khaki uniforms are the same ones I wear now, and this is my chance to be one of those brave, heroic men sweeping the seas clear of hazards to Freedom. And this is a war of right versus wrong; of communist invaders in democratic South Vietnam. Why would I NOT volunteer?"

Well, there WAS a possibility of injury or death. Furthermore, the kinds of duties offered were rather different from anything I would expect to fall into my career path, and volunteering would require my early departure from this Chief Engineer job. Not especially career enhancing. Anti-war protesters? Screw those long-haired Peaceniks; they must not be allowed to subvert our nation.

There was just one compelling counter-argument, to my mind - I had met a girl, and it was serious. However, that relationship blossomed at a time when it was too late to call off my adventure - my letter was already in.

So, off I went. Two months of Pre-Deployment and Counter-Insurgency school, for instruction in Vietnamese culture and language, weapons familiarization, and a week of SERE school - "Survival, Evasion, Resistance and Escape". A week with no food, little rest, and mistreatment by a bunch of make-believe enemies with a POW camp.

The late-blooming romance led to a quick wedding, a brief honeymoon, and a move into a furnished apartment before I reported back for a very long plane ride: San Francisco to Saigon, via Alaska for refueling. It was March, 1968 – just days after the "Tet Offensive". Arrive Saigon at 7:00 A.M. (9:00

P.M. the previous day, according to our body clocks). Hot and dusty. Badly damaged airport terminal. Climb on an Air Force bus with heavy window screening (to deflect grenades). Driver sporting a sawed-off .30 cal carbine on the dash. Lengthy drive through a teeming Oriental city, sandbag bunkers everywhere. Welcome to My War!

My assignment was to guide and assist the commanding officer of a 112 foot patrol craft in what turned out to be a Coast Guard like function. We were deployed for 45 days at a time to patrol the coastline, preventing clandestine resupply via fishing or other boats. During deployment our vessel operated out of one of four Coastal Zone HQ's; between deployments we returned to Saigon for repair and resupply. Never in my year did I feel personally threatened. There were unfocused neighborhood threats - sometimes at night you could hear mortar rounds fired and landing within a block, but the target was a police station down the street. Rarely, explosives were found attached to the undercarriage of vehicles used to transport servicemen - but never one I was in.

Once during a deployment to the remote Fourth Coastal Zone, centered on Phu Quoc Island (near the Cambodian border on the southwest tip of Vietnam) we had the misfortune of finding a reef. The vessel sat, for seven days, within a few hundred feet of jungle-covered shoreline. We remained partly submerged until a U.S. salvage ship arrived to extract, patch, and tow us back to Saigon.

My year was lightened by a single, seven day R&R in Hawaii with my bride. We spent Christmas together, despite everything. In those Pre-TSA days, she even brought a small Christmas Tree to brighten our hotel room. Then, in March 1969, I climbed aboard another chartered airliner for the trip home. It seemed ever so much quicker than the trip west! Landing in San Francisco, I transferred immediately to another plane, and continued on to San Diego where I was greeted by my wife.

Protesters? Never saw one.

Looking back, my risk was rewarded. I am proud to have served my country at a difficult time. My service was never a black mark in my world - the armed forces were well insulated from the haters. Five years later I was shamed by my country's failure to live up to promises of support once we departed, but by then I was moving on in my career, which was NOT damaged by my year as an advisor. Today, fifty years afterwards, once-Communist Vietnam seems closer to an ally than an enemy. I don't know what happened to the South Vietnamese people I met, but do know many have been living in the USA for years.

Sometimes, now, I wonder - what was it all about?

PARENTHOOD

14

Character

Introduction:

The Essay presented here (in a few pages) is totally bogus! (Actually, you may decide it isn't totally bogus, after all – you must be the judge!) The assignment felt like more of a Challenge than a routine assignment. Therefore, I have included the Assignment for the reader's orientation. The essay was mine, NOT the creation of my daughter – in other words, *Don't Blame Her!*

The Challenge:

Assignment for Week Seven by Jack Galvin - April 6, 2015

We're down to two more shots, two more chances for insight into aspects of our lives.

I hope the assignments have become important to you. I say become, because if you're like me, you are not sure how important each one is at first, or will be, until you think about it a bit and then a layer gets peeled back by the thinking and then a process begins: recalling and then writing, then the "aha" moment when it all becomes important. Some topics may have struck just the right nerve for you to play back that memory tape and write about it in such a way that you surprised yourself.

This assignment will be, as you will soon see, perhaps the most challenging yet. But that's as it should be. With only two more to go, the assignments ought to demand more of us--more of our memory, our insight, and our candid, vivid, writing.

So here goes.

James Hillman in his *The Force of Character* and the *Lasting Life* examines a host of aspects that contribute to each of our unique sense of character. Near the beginning of the book he asks, "What is the "it:" that persists and endures? What could possibly last through all the events of a long life, remaining constant from start to finish? Neither our bodies or our minds stay the same: they cannot avoid change." Then he answers his own question. "What does seem to hold true all along and to the end is an enduring psychological component that marks you as different from all others: your individual *character*.

Then, in one of his last chapters, he enumerates what he considers to be the virtues of character. I want you to consider three of them here:

1. Character is perceptible as an image. It is on display as style, habit, disposition, habit carriage, mien, presence. The face reveals character and appeals to character. As image, character must be imagined as well as perceived.

2. Character has always been distinguished from talent, skills, gifts, and measurable abilities. It can be lamed by deficiencies and trapped in fixations, all the while talents and skills exhibit brilliance. Character does not yield to standardized measures of performance. The characteristic uniqueness of a style eludes analysis.

3. Character also eludes the moral clamp. It reveals itself not in the morality of behavior but in its style. Character traits include vices and virtues. They do not define character. Character defines them. Perseverance or loyalty may instigate a criminal act as

> well as a just one. Friendship may motivate revenge as well as self-sacrifice. The imaginative scope of character cannot be pressed into an ethical definition without perverting its nature and sterilizing its fecundity.

Now to your assignment.

I want you to consider a time or even a moment in your life that demanded great endurance, tenacity, or strength from you, a situation that spoke clearly of your character. I want you to write about that time as vividly as you can, using the techniques of scene development, plot and theme that we have talked about.

But here's the catch. I want you to write about the situation **from another point of view**, from the point of view of someone who would know about how you responded to that situation. In other words write about you from the point of view of someone who watched you go through the experience or knew the situation intimately enough to have the insight to write about your character.

This is another exercise in writing but it is also an opportunity for us to see ourselves in tough times as another might have seen us. It allows us to examine our character as we think we present ourselves to others. It is neither falsely modest, nor falsely self-aggrandizing. It is just the way we were as we think someone else may have observed us.

This assignment will be a challenge to all of us, first to find a moment or situation that showed our character, then to imagine another person's perspective of us.

Good luck.

The Response:

The following pages were originally printed as "By Elizabeth Heins" – but I have decided to delete that tag, to lessen confusion.

"Character"

I've been asked to tell you about my father, Commander Raymond Rice Heins, USN (Ret). He served 20 years in the Navy, raised four of us kids, and is currently living the retired life in Rhode Island. Specifically, I was asked about his "character".

Can character be summed up in a single word? Let's try one: Stolid. The word reminds me of "solid", but is defined as "Calm, dependable, and showing little emotion or animation". Synonyms include "dependable" (that's good, right?) but also "unimaginative, dull" (not so good). Antonyms include "emotional, lively, imaginative".

All his life, I suspect my Dad has admired those antonymic qualities, while knowing deep down that "stolid" probably described him much better.

When did it begin? He says one of his earliest memories is of visiting an old lady who lived directly across the street from his house. He sat politely in her parlor, on an uncomfortably formal chair with scratchy upholstering. He already knew the lady kept candy ready at hand, and at the end of his visit, she would give him a candy bar from the soft, wrinkled paper bag in which she kept it. His mother nailed down the lesson when she explained that Mrs. Sawyer thought he was a wonderful child because he never asked for candy. So the candy was a reward for restraint. Good things came to those who wait. Stolid.

Another story he has related a few times: Emotional behavior didn't serve him well in childhood; as a rule, it only led to unpleasantness or punishment. An exception proved the rule: He was about four, with his father at a Knights of Columbus carnival. A game of chance booth offered as a prize a toy fire engine, maybe 14 inches long. It was a hook and ladder truck. He wanted that! Screaming and crying, he demanded that! His father couldn't seem to win it, so more screaming and crying. Finally, my Grandpa somehow acquired it - probably by outright purchase. Strangely, he says, he (Dad) never liked the toy, and felt ashamed of his tantrum whenever he did play with it. Stolid would have been better.

He observed that students who never caused trouble were fondly regarded by teachers, and he always wanted to please adults. Later, he always wanted to please those in authority - teachers, coaches, military superiors.

With peers and subordinates, he wanted them to see him as a dispassionate, fair, professional leader. Stolid, you might say. Not inspirational, maybe, but dependable.

So how did it work out? I would say there were mixed results.

Professionally, he never made it to the top. It became clear that he probably would not screen for Captain, let alone Admiral, so he submitted his retirement letter before the Captain selection board met. He did, however, make it to Commander, serving on two different ships as a department head, and on a third as Executive Officer. He also was the good student, earning a Master's degree in Computer Systems Management, which led to two successful tours of duty in related fields, and a ten-year management stint after retirement in computer systems development. In all those jobs, he went for dependable and unflappable, rather than excitable and imaginative. Stolid.

In private life, stolidity may not be ideal. Touchy-feely, with lots of hugs, probably works better than remote but dependable, with both children and spouses. Unfortunately, stolid does as stolid is; it's not so easy to shift character between home and world. I don't mean to complain; he was never a harsh disciplinarian - In fact, there is a secret core of kindness, best recognized by children and animals. God knows he was a soft touch if we played our cards right. But somehow I think we could have been a more close-knit and loving family, with a little more emotion and a little less cool stoicism.

Character can be inherited, I guess. Maybe the whole stolid thing is part of our Germanic heritage. Certainly everyone in the family exhibits the good student part, and has pursued a post-graduate degree - even the sister who blew off college on her first go-round. Before her MBA, she obtained a Bachelor's Degree the hard way - night school. And she did that on the quiet, never even advising anyone in the family until she was almost through, after some six or eight years of part time work. Talk about stolid! Anyway, none of us children married, until baby John did last year, when he was already 30 years old. Now there are five Masters' degrees in the family - six counting the new sister-in law - but just the one marriage (besides our parents). I think the family has lots of self-discipline, focus, and persistence, but not a lot of spontaneity, imagination, or sensitivity. Sometimes I think we're just a stolid mass!

15

Michael - the Struggle

Our first-born son did not hesitate to leap into our life. Born in January, 1970, he must have been counting on our inexperience as parents. He was aided and abetted by the Army doctor who oversaw his birth, at the Fort Ord Army hospital, near Monterey California. Neither the doctor nor any member of the staff gave us any hint that our son was not perfect in every way. And yet…

We might have known, if we were more experienced. The almond eyes, stubby hands, slightly flaccid muscle tone, all symptoms of Mongolism, to use the horrible, racist, outdated terminology which still lingered then.

Grandmother knew. A nurse for more than twenty years, she must have known as soon as she saw her first-born grandchild. She arrived the day mother and son arrived home from the hospital. Grandpa called, from their home in Pomona, and I picked up the extension to be part of the conversation. But before I said a word, I heard him finish his question:

"Is there any hope?"

"No.", she replied curtly. Just one terrible word. I didn't speak, waiting for their conversation to end so I could replace the receiver quietly. I didn't need any more confirmation, as I had already begun to suspect. No hope – my son would never be a normal human being.

Some of my fellow-students at the Naval Postgraduate

School were also experiencing first parenthood. I tried to avoid discussions whenever possible. "Both are doing well", I would say. But secretly I envied them.

Michael's development was slow – he was late to walk, late to speak, and when he did begin talking it was hard for him to make the sounds he heard, and he was hard to understand. We began to find the resources the birth hospital should have provided, and soon learned that our son was "just one of those things". One in 600 newborns, they told us, had Down Syndrome. We even had chromosome studies done (courtesy of military medicine) showing that we parents had no more likelihood than any other couple of having a child with Down Syndrome in the future.

So, as soon as practical, a sister was delivered. We did have a test performed during pregnancy; God only knows what we would have done if the test indicated another Down child - I don't think we could have countenanced an abortion. But the test was inconclusive. "Either it's a normal girl, or we picked up tissue from the mother." Thanks for nothing.

Elizabeth was born in August of the year following Michael's birth. She was tiny, but perfect. Still, her mother was burdened with two infants, while I sailed the seas with a newly commissioned ship, USS *PONCE* (LPD 15). The ship spent even more time underway than most, out of touch with families. We had systems testing to conduct, then departed for a Sixth Fleet deployment. Meanwhile Virginia and two non-walking children were deposited in a Virginia Beach apartment, isolated from family, husband, and even other ship's wives.

This rocky start to our married life did improve, some, when we moved into government quarters at Little Creek, located between Virginia Beach and Norfolk. The house was small, but was surrounded by grass and towering oak trees. There was a wonderful sun room, which served as an oversized play pen – filled with toys, always warm and usually sunny. Mike and

Beth were quite content to be there with each other. Mike finally began to walk pretty well, about the same time Beth did. Although his speech still was difficult, he was able to make his needs known. Perhaps best of all, there were friends in the other little white houses, so that Virginia had some social outlet with other wardroom wives when the ship was out.

This quiet interlude came to an end when I was transferred to USS *VOGELGESANG* (DD 862), home ported in Newport. We were in navy housing again, but under less happy circumstances. The Navy was in the process of evacuating from Newport. Most ships were gone, along with their crews. We moved into a neighborhood in which less than half the residences were filled. It was while we lived here, nominally in Portsmouth, that Michael began school. He went to a Special Education classroom in Melville School, and thus began a new phase in our lives.

After a single year, we moved again, to Herndon Virginia, a suburb of Washington. Michael attended Oak Grove School in nearby Fairfax. In many ways it was the best arrangement ever. The entire school was dedicated to mentally challenged children. The teachers were devoted, the children were loved and protected. The Christmas pageant gave each child a chance to be a snowflake, and every one was adorable. I became quite sure that this was the very best way to educate our child.

Alas, other views challenged this belief. "These children need to see the real world." "They will never learn properly while they are isolated." "Other children need to be exposed to them, as well." Less welcome but perhaps most telling, "This is the most expensive way to educate – our town cannot afford to dedicate a school to Special Education!"

After three years in this cocoon, we returned to Newport, when I was assigned to the Naval War College – as a student the first year, than on the staff for four years. During my student year, we lived at Fort Adams naval housing, so Michael attended Newport's Mumford School. It was here that Michael was so rudely introduced to academic rigor – his teacher, a portly, bearded 40-year old man, forced Michael and his classmates to learn to read and write! At age eight, I was sure he was not capable of such learning, and yet learn he did. Reading for pleasure never was in the cards – he just wasn't interested – but he was able to puzzle out words, and we soon realized that he could read what he wanted to – including the "MEN" sign on a restroom, and "Cheeseburger" on a restaurant menu.

After just one year at Mumford, we moved to Middletown, and Michael attended elementary, middle, and high schools. Nowhere did he encounter the loving isolation of Oak Grove, nor the challenging and transformative curriculum he had found at Mumford, but he was complying with state requirements, and being exposed to reading, writing, and arithmetic – if you didn't look too closely. He was also being exposed to other students, in a controlled environment. There were no serious problems, and he progressed toward adulthood.

Michael's out-of-school activities gradually increased. He joined a Boy Scout Troop, as a "Special Scout" (along with his father – I became an Assistant Scoutmaster, so I could be there with him). He became an altar server at St. Joseph Church. He began to swim at the YMCA on Friday evenings in a program called "Adapted Aquatics". He took karate lessons, initially with a class of similar-age kids, then by himself. And he was a regular participant in Special Olympics two weekends a year. Around town, our family would occasionally encounter young people who knew Michael, and greeted him in public. He seemed to know them, also; every one of them was named "Dude".

Mike remained in high school until after his 21st birthday. As it developed, he had enough credits to graduate with a diploma in June, 1991. Also at that time, he transitioned into services offered by the Maher Center – a non-profit "Rehabilitation Center" which is, in fact, a lifetime support center for mentally handicapped Rhode Islanders.

Despite all the hopeful signs of Michael's maturing development, there were shortcomings. His speech never improved noticeably, despite some ineffective speech therapy he received during part of his schooling. Most seriously, he began to display a temper, on occasion.

The first time I remember this temper, we were on a Scouting weekend. We were sitting on a log, along with another father and son, when something displeased Michael. Without warning he struck me, hard, on the ear. Unprepared and unsure how to handle this, I tried to make him feel guilty: "Mike! You hurt me! Why did you hit me?" I also moved out of reach, but never did figure out what prompted the swat.

Not long after Michael moved from high school to the Maher Center; the center was awarded a government contract to staff the dining hall on the Navy base. I worried about his ability to keep up in such a demanding environment. Those sailors weren't part of an experiment – they would want their food, without delays. Michael and his peers would be cleaning tables and floors, loading and clearing dishwashers, loading table condiments, and tending the food line. Professional cooks would be preparing the food, but the operation couldn't work without Michael and his friends. It proved to be fast paced and demanding employment, but happily it worked out well. For five years, Mike would rise at 4:15, dress and eat, and be ready for the transportation van sometime between 4:50 and 5:00 A.M. His mother and I took turns getting up with him, returning to bed after he departed.

This routine worked out surprisingly well. Michael was earning real money, at $5.00 per hour. Although the money meant little or nothing to him, he seemed proud to be holding a meaningful job. But sporadically, we would receive a very unwelcome call, usually around 6:00 A.M. Mike's temper had flared up, and he had acted out, and he needed to go home. Once he toppled an entire five-foot high cart of dishes. Another time he pushed over a milk dispenser. Sometimes he took a swing at his supervisor, who just happened to be a 6'2" mountain of a man, at least 300 pounds, with the sweetest disposition you could hope for.

Michael never could explain his outbursts, but usually was extremely remorseful. Sometimes, when his explosion occurred at home, we witnessed remorse interjected with additional swings. He would pass through a red-faced anger to a quivering, white faced rage. He would lash out, then sob, "I'm sorry", before another attempt to hit. These assaults were directed at his mother, his father, or even occasionally at a sister or brother. Our response was to send him to his room for an extended time out. Sometimes this resulted in damaged furniture or appliances, as well.

Finally we obtained a referral to a psychiatrist. The doctor was a woman experienced in dealing with the mentally handicapped, but she seemed woefully ineffective at dealing with Michael. He was always perfectly friendly and cheerful when visiting her, and she seemed to have no conception of how to communicate with him. Ultimately, based on our detailed log of at least ten incidents, she pronounced a diagnosis of "Intermittent Explosive Disorder." We could have saved everyone some time – we already knew that much!

Things did not improve. Michael was suspended from work when he struck another ("normal") worker in the head with a plate. When I arrived after the early morning call, I was advised the base police had been called in by the victim. Two officers had taken statements, and collected names of principals, witnesses, and supervisors. Recognizing that Michael no longer posed any sort of threat, and that I would be removing him to home, they departed. No formal charges were lodged, but this was the end of Michael's tenure at the galley. He was given an indefinite suspension, and then, in October 2006, he was terminated. This was probably the lowest point of Michael's life. He had lost a job he greatly valued. He was repeatedly harming his friends and family. On top of that, and possibly related, was the fact that his younger brother had finished college and moved away. Both his sisters lived in distant cities. Only Mike remained behind, with no prospects of moving on with his life.

We had a standing request for Michael to move into a Group Home. Clearly he would always require supervision, but we were increasingly convinced that he needed a change of scenery. The explosive outbursts were unacceptable to us, and could eventually result in serious harm to someone. On this basis, he was accepted for assignment to a group home in Barrington, Rhode Island, in October, 2007.

So now he had two homes, and he was always happy to move from one to the other. He was happy to come visit his parents in Middletown, and happy to return to Barrington. His housemates there were invariably happy to welcome him back "home", and one or two of them would follow Michael around while he re-settled.

At that point I saw increased self confidence in Michael. He seemed to be the master of every situation. When he was in Middletown for a Sunday, he loved to go with us to St. Joseph Church, where he glad-handed as though he were running for office. He still seems comfortable calling anyone "Dude", but is equally comfortable with unspoken greetings. At work (for a time the Maher Center itself, either in the workshop, or as a part-time janitor, or delivering sandwiches), his supervisors praised his cooperative spirit. He had become a happy and confident adult. A Triumph? You bet! At least for now...

Time moved on. In 2021 we (both Michael's parents) eventually moved into an Assisted Living home, and sold our house in Middletown. Michael, now over 50 years of age, remains in the Barrington Group Home, and now works in janitorial services on the navy base. He is still proud to be living a productive life, and still has the occasional (now very rare) episode of angry outburst. Once again, I think he is settled for good. (At least for now…) Below are photos of Michael at 18 months, and Michael at 50.

16

You Are My Sunshine

Our children began to arrive in January, 1970. The second followed just nineteen months later, the third in 1975. The youngest was born in July, 1983. I was happy to be a hands-on father from the beginning. There was no diaper too smelly, no mess too disgusting.

In the children's earliest infancy, I was happy to leave the bathing to my spouse. Those tiny but noisy creatures fit nicely into the kitchen sink, and there they were bathed. At first. Eventually, though, they would become too large for that, and (happily) they learned to sit up about the same time. Soon they were ready for the bath tub.

Naturally, my wife made the call, and began bathing the children in a tub. But I was ready to take my turn assisting with the evening bath. At first terrified by those slippery bodies, I quickly learned how to provide assistance in remaining vertical. Once both I and the child of the year came to trust that I would not drown him or her, we began to enjoy the time.

Usually there were floating toys, of one type or another. "Rubber" ducks (actually plastic, of course) were always fun, but boats were welcomed, as well. After all, I was a navy man, and we could pretend that these were ocean-going vessels - although not very sea worthy. Capsizing was frequent, and sinkings not unknown.

Between lathering and rinsing, we began to introduce songs.

A frequent and favorite effort was "You are my sunshine":

You are my sunshine,
My only sunshine;
You make me happy,
When skies are gray.
You'll never know, dear,
How much I love you.
Please don't take my sunshine away.

I knew there were probably more verses, maybe many others. But these few lines were easy to remember, and expressed my feelings very well. The bather would join in lustily. At least one child, for a while, couldn't seem to master the trick of changing the tone of the notes, so she would just sing certain words more loudly than others.

I thought at the time that this was a traditional folk song in the classic sense, with no identifiable origin. Later I learned (thanks, Internet) that it had an extensive back-story. Written around 1939, the song was very popular in the early 1940's, first in a version by Jimmie Davis, who was both a Country Music performer and governor of Louisiana. In later years, it was recorded by an astounding range of singers, including Gene Autry, Bing Crosby, Lawrence Welk, Ike and Tina Turner, Aretha Franklin, and Andy Williams. (Imagine: Ike and Tina Turner, and Andy Williams?!)

The song was familiar to me from my own childhood. My mother would sing it, along with a larger repertoire which included "Clang clang clang went the trolly", "Buffalo gal", and "On Top of Old Smoky". Some of these were of recent vintage, but others old. Some must have lingered from days of slavery - like "Jimmy Crack Corn" (...the Massa's gone away), and "Short'nin' Bread" (...stole the little gal who makes short'nin' bread.) My mother was born in rural Georgia in 1901, 36 years after the end of the Civil War, and long before the days of radio, electricity or indoor plumbing. She told us children of her recollection of a visit to her school by a Confederate veteran. I have no doubt that plenty of songs retained popularity more than 40 years - after all, Rock and Roll today is 60 years old!

"You are my Sunshine" was unique, I think, in that it so strongly affirmed the love of the singer. Having children was the high point of my life, and I found it hard to express that fact to them. When I sang our song, I did mean it, from the bottom of my heart.

In later years, we went through lots of nursery rhymes, Rafi, alphabet songs, "Itsy Bitty Spider", Old McDonald. But in retrospect the one song I remember best, and which still tugs at my heartstrings in a pleasant way, is "You are my sunshine". I can only hope that someday I will have the opportunity to sing it with a new generation.

17

You Can Go Home Again

Arriving home from Logan Airport. Easing down the street, in my two-day old car, deep snow everywhere. Thank goodness, the snowfall eased up earlier today, and the snow plows came through. I used the snow blower on the driveway, opening a nice, straight path between three-foot high walls of snow. Holy crap – is that wide enough to get into??? I brake, and began the turn. The car cruises serenely ahead, responding only in part to my command. Bump! With a gentle impact, the slide ends. Beloved daughter Beth gazes serenely from the passenger seat, and says: "What now?"

Well, it's obvious; we back away from the snow bank, and aim more carefully. Only the car does not cooperate. The wheels turn, but we move not an inch. The street grade, though not extreme, is enough to resist the meager traction available to the tires. If the car moves at all, it is only to more thoroughly block the street. Now no one could get by without damaging my brand-new car, if they were silly enough to try.

It is December 20, 2009, shortly after the early dark. Beloved second daughter Mary arrived earlier, via T.F. Green. Wife, daughters, and I try heroically to push the car up our Sisyphean hillside, but no dice. I pull out my AAA card, and beloved daughter Beth memorializes the scene. Mom and Mary still pretending to push, I holding aloft the magic card.

Mom and Mary still pretending to push, I holding aloft the magic card.

Dinner was waiting inside; a seasonal feast featuring root vegetables, savory meat, lovely soup. So I called AAA ("I am blocking traffic, so please hurry"), and we relocated to the dining room, keeping an eye out the front window. Christmas began, although not everyone was able to be there. For the first time, one of the family would miss the Holiday – John was in Afghanistan. Michael, now residing in a group home in Barrington, would be picked up tomorrow, so his sisters could visit him in his home.

Even without a full house, it was still Christmas. The tree was standing in its focal point of the living room, with lights already strung. Some decorations ("ours", as opposed to those of the children) were on the tree. Stockings were hung with care, the Crèche was on display. More important, the entire family, insofar as possible, was present. Even without John, it felt like Christmas, although his absence was palpable.

The tow truck arrived before I finished my meal. In moments, it seemed, the operator attached a chain to the undercarriage, and pulled the car free from the snow bank. I then drove it into the drive and the garage, and returned to the feast. Afterwards, Beth and Mary added their decorations to the tree, and we shared a glass of nog. We toasted John, and looked forward to his Christmas phone call. But John's decorations were not hung.

2009 was the first year a good snow preceded Christmas (by just a few days) in my memory. In the Tennessee of my childhood, we could count on two or three snow storms per winter, but they never started until January. Generally, that seems to be the same pattern in Rhode Island now, although we can occasionally muster the freak snow storm much earlier than Christmas.

So what makes it Christmas? Snow helps, but is not essential. Cold weather? No, one excellent Christmas was spent with my new wife in Hawaii, after nearly ten months in Vietnam. And in Tennessee, I delivered newspapers in shirtsleeves (my new gray turtleneck) on Christmas afternoon. Music? I do enjoy Christmas music, and it does signify the season, even though it begins to fray at the edges all too soon. (Christmas music is totally gone from the air waves by December 26th.)

Religion? I recognize the "true meaning" of Christmas, and try to keep it in mind, but really, theologically, Christmas is not a big deal. The date was chosen to co-opt a pagan celebration of the winter solstice, and the crucifixion and resurrection are far more important from a religious point of view.

Presents? The wrappings are lovely, sparkling in gold, shining in red and green, sporting the colorful depictions of Santa, reindeer, and blue, snowy images of Bethlehem (which probably has never seen snow). But the presents are just a side show.

Maybe it's the lights – all the colors of the rainbow, buried in soft fir needles of a tree in the living room and peeking from the windows of other houses; or the white lights decorating city streets for weeks in advance, and lighting many houses in my neighborhood for the last weeks before Christmas. I go to quite a bit of trouble to install white lights in the small trees of my front yard, and to plug them into a timer, so our house will add to the season.

But more than anything else, it is "Home" and the gathering. My children have returned home for Christmas each year of their lives, with the sole exception of Afghanistan John.

When I was John's age, serving on ships, I frequently had to choose between Christmas and New Year's - someone had to be on board at all times! In addition, one Christmas was spent in Hong Kong. (I sent telegrams home on the afternoon of Christmas day, and they arrived on Christmas Eve!) But even when I could only arrive on December 29th or 30th, I had the sense of being Home for Christmas. My parents had the tree up, and decorations around, each of those years.

In 1968, having married, things did change. No longer was "Home" where my parents were. It came to be wherever my spouse was, and later where she and the children were. Since none of my children are married, that remains true, in my heart, and to date the children have played along. If and when they marry, I realize things will change again. There will be more than one "Home" then. Quite clearly, the truth is, home is where your heart is. There can be a different "Home" for every member of the family, but I will always believe that each of us can, indeed, go "Home" again!

PROFESSIONAL LIFE

18

An Ending and a Beginning

Told by Elizabeth Heins

NOTE: The following essay resulted from a class assignment - "to write in another voice." (Beth didn't really have any role in writing it – my apologies to you, Beth!)

My Dad had to make a momentous decision around 1994. He had been working for Computer Sciences Corporation for ten years, responsible for the maintenance contract on a large-scale computer-based war gaming system. This system was installed in three locations: Newport, at the Naval War College; Dam Neck, Virginia at the Tactical Training Group, Atlantic; and San Diego, at the Tactical Training Group Pacific. In addition, a partial development system, where future versions were created, was located at Moorestown, New Jersey (a distant suburb of Philadelphia).

In the last year, he had been promoted to manage the future systems development, as well as maintenance of the current, fielded system. This wider management responsibility required him to work in Moorestown. He commuted on a weekly basis for almost the entire year.

My father began to face a choice: after nine months of weekly commuting, and he needed to make a decision. Did his future lie with the company and Moorestown, or would he cast off this ten-year employment in order to keep the family in a familiar location?

Mom certainly had an equal vote in this choice, if she chose to cast one, but she insisted he should feel free to make an unconstrained decision. While the very idea of moving to New Jersey was a little scary, I knew there were livable suburbs in the Moorestown area. Besides, at age 22, I was well into my four years at Stonehill College, and already realized I probably would not live much longer in the Rhode Island area.

For Dad, it was a choice between continuing a satisfying and challenging career, or starting all over again, in his mid-50's, at something new. With a navy pension, and savings from the CSC years, he had some flexibility of choice. But I think the heaviest weight came from the idea of moving the family to a new place, versus remaining settled in Rhode Island.

Up until arriving at the Naval War College for duty in 1978, our family moved A LOT. I was just seven years old at that point. My disabled (Down syndrome) brother, Michael, was a year older. My sister Mary was actually born in Rhode Island in 1975, but we moved almost immediately, and lived in northern Virginia from 1975 to 1978. So Mary was just three when we arrived at Newport. John, the youngest, was born in 1983. So here we were in 1994, settled in Rhode Island for 16 years. Was my Dad too timid to take the move to New Jersey? Or was he giving such weight to the desirability of stability, that he was willing to give up a positive career move?

He seemed to love living in Rhode Island. Sometimes I wonder why, considering the weather and the economy. It's not as if he was a yachtsman or a surfer! Maybe it had everything to do with the children. Michael (23 in the year of the decision) was very well placed in Rhode Island. He was supported at the Maher Center in Middletown, through which he was employed at the Navy base as a food service worker. Mike was clearly very proud of the fact that he was doing an important job – even though it meant rising at 4:30 every morning. I was in college just 45 minutes away, in North Easton Massachusetts. Mary was in transition, more interested in establishing independence than in college, but she lived in the area. John, the youngest, was a student at St. Philomena school, and very happy there. A move would have been easy enough for the family to tolerate, except for Michael. It was hard to imagine a placement anywhere else which would be as satisfactory – and both my parents did explore the options in New Jersey.

So finally, the die was cast. Dad notified CSC that he was not willing to relocate. CSC sent him a layoff letter. He had to recommend a successor from the staff still in place. In June, 1994, he left the Naval War College facility for the last time and began a new life.

19

A Boss Under Pressure

NOTE: Yet another assignment to write about myself in the voice of another person. (No offense, Rocky!).

BEGIN AT THE beginning, they always say. OK, I've been working on this large-scale computer system for a little over five years - joined CSC (the developer) in 1989, when they had a vacancy on a software maintenance team. I like to be called Rocky, because my last name is Rhodes, and it beats "Dusty" - let alone "Roger", which my Mom laid on me. For seven years before that (OK, so I didn't begin at the beginning, after all), I worked in the Gaming Systems department as a Navy enlisted guy. I experienced war games with no computer support, and games with the NEGS - Naval Electronic Gaming System. To tell you the truth, I don't think the system added much to the games, but it did require sailors who could operate the terminals. Over time, I became pretty good at that - also preparing the games in advance, and making changes to the database of ship and aircraft capabilities - weapon systems, range, speeds, sensors, detection range. That kind of stuff.

Anyways, I went to work for CSC as a tester. The boss in Newport is Ray Heins - early 50's, one-time head (as a Navy Commander) of that Gaming Systems Department where I worked as a sailor. I didn't know much about him then - he was my boss's boss's boss - and still don't know much about him. Now he's retired from the Navy, but is still a close-hold kind of guy - doesn't give much away, if you know what I mean.

Lately, we've begun to hear rumors of change in the air. Not that we hear the early scoop from our Leader. He's addicted to memos, it seems to me. Never tells us in person if he can write a memo to tell us. But some of the guys working in New Jersey are a little more forthcoming (they do development and testing on the next generation system, while we work on the currently delivered version).

The first guy I heard from was Cooper in NJ: "Looks like SPAWAR is running into funding reductions. Don't be surprised if some of us in Jersey will be affected - and maybe Newport staffing, as well." (SPAWAR would be the Space and Naval Warfare Systems Command - the government, for our purposes.)

Cooper gets his information second hand, through the bean counters, but he doesn't sound all that worried. I'm thinking he feels pretty secure in his niche. And besides that, CSC has lots of other projects going in NJ, so staff reductions from NEGS could probably be picked up on something else. But that's not at all true in Newport; laid off would be laid off, unless you are a programming whiz - which I ain't. I think Cooper just likes feeling superior, which is why he's so happy to spread the bad news.

After about a week of rumors, Our fearless leader called a company meeting.

"I guess you've all heard the scuttlebutt", he started (he likes to use Navy slang). "I will tell you what I can. It's true Navy funding has been tentatively reduced for the upcoming fiscal year. The SDF (development site in NJ) will be supported as the primary effort; the operational site here will continue to operate, but will receive less contractor support. In other words, we will be taking most of the hit. The new fiscal year doesn't start for almost six months, so we have time to work out a reduction plan."

Oh Goody, I thought. We can select our grave sites. But there was more:

"I will be one of the earlier reductions. I expect to be released in about two months. I will have to recommend someone to take over the reduced set of tasks still required of management - like time cards and assigning duties to whoever remains on site. And, of course, I will have to draw up a plan for personnel layoffs. I expect all of you will be scouting around for new jobs, if you aren't already. Please talk to me as you move forward, so I can minimize involuntary terminations."

And so began a very awkward month. I wondered how it had happened that Heins was identified as an early departure. He almost surely could have moved to NJ and remained with the company in another assignment. Brenda, our combination office mother and secretary, claimed he volunteered to be released up front because his salary drain was highest of our staff. Also, she said, he didn't want to move his family to New Jersey. "Why not?" I wondered. But she didn't venture a guess.

I quickly realized the choice of new manager would be significant to me, personally. Ours was already a small staff - twelve people - but if it got much smaller, the manager and every remaining worker would be constantly rubbing elbows. I would probably have daily interaction with the new guy.

I decided there were two obvious choices - Leonard (Zeke) Wellington, and Mike Schubert. Both were old dogs, like Heins. Zeke had spent 20 or so years in the Navy - as some kind of pilot. Mike had spent lots of time in sales, and seemed like a used car salesman to me. He didn't inspire confidence, in my humble opinion.

Both these guys had experience in using NEGS. They could set up games, man the computer consoles, and run post-game analysis. Neither was a programmer, so they couldn't fix a system failure, but they could gather the information that would help a programmer develop a fix. Zeke was a stand-up guy, with strong religious beliefs and a long marriage. Shubert was a womanizer - everyone knew it - and was wearing through his third marriage. He was a loud, extroverted, back slapper, a good man at a party. Zeke, like Heins, was much less comfortable in that back-slapping role.

I was very interested in seeing how the boss would decide on his replacement. I knew he had worked with Zeke for a number of years - both of them were war gaming umpires in the pre-NEGS days, and both had watched the system take shape over its development - before either of them were associated with CSC.

About a week later, the word was out. The winner was... Shubert the reprobate. As usual, Mr. Memo floated a memorandum saying for the record that he anticipated Mr. Shubert would be named the site manager upon his (Heins's) departure. I felt close enough to Zeke to ask him in private how's come his friend Heins chose Shubert for this job. He said he had asked that exact question in a private meeting. The answer?

"Well, Ray was pretty mealy-mouthed about it, but I think he feels Schubert will have a better chance of keeping the site in business than I will." said Zeke. "I strongly disagree, but he says Mike is better at communicating. I tried to tell him, as a friend, that no one trusts Schubert, and therefore they aren't very likely to believe anything he tells them, but he just can't see that. He seems to believe that because Schubert can write a better memo, he can explain the importance of the job being done here. He actually apologized for passing me over."

Clearly Zeke was still not convinced, and he was not very happy. So I was left with these questions: Did the boss cave in to NJ pressure? I'm pretty sure Schubert's drinking buddy down there had the final say, anyway. But why did he not elect to move to NJ? I was still curious about that, and asked Zeke.

"I think I can answer that one. You know he has a son with Down syndrome, don't you?"

I didn't, but so what?

"He (the son) is very well placed here in Rhode Island. There are a lot more opportunities here than in New Jersey for employment and social connections - and he's already plugged into that network. Also, Ray has three other children, all in local schools, plus his wife works at one of the local churches. I think he decided the upheaval wouldn't be worth it."

So, I concluded he did knuckle under to pressure from the home front, so maybe he also knuckled to pressure from NJ to name Schubert the new manager. I am left with mixed feelings, for sure.

20

Computer Quest

"I want to work with computers"

My first exposure to computers was in 1962, when I received about four sessions on computer systems as part of a naval science course at the Naval Academy. We had to prepare a program which would calculate intercepts against an orbiting weapon system. This was just five years after the Sputnik surprise of October, 1957. The United States, caught flat-footed in a space race, recognized that computers would play a huge role in our space program. I found that intriguing, and in short order, realized that I really wanted to be around computers.

Fifteen years later, in 1977, my fascination with computers took on another dimension, as personal computers became a reality. That was the year of the Apple II, the Radio Shack TRS-80, the Commodore PET, and Heathkit "build it yourself" computers; I knew I had to own a computer! But that fever was still far in the future in 1962.

"Home Computers" had been anticipated long before 1977. Note the view of one of them from 1954 on the following page.

To scratch my itch to work with computers, obviously I would need some heavy duty education, and some luck. Computers did exist in 1962, but they were special-purpose, industrial scale machines.

"Scientist from the RAND Corporation created this model to illustrate how a "home computer" could look like in the year 2004. However the needed technology will not be economically feasible for the average home. Also the scientists readily admit that the computer will require not yet invented technology to actually work, but 50 years from now scientific progress is expected to solve these problems. With Teletype interface and the Fortran language, the computer will be easy to use."

Each was served by acolytes who changed tapes, fed in punched cards, tended printers. Careful management was needed to make sure the beasts were kept busy and productive. Entry into this priesthood required talent, education, and being in the right place at the right time.

In 1977, as the scale and price of computers began to plunge, a new need began to emerge; new ways of working – new programs – had to be developed to let more people do more things with them. "Word Processing" would assist, then replace, typewriters. "Spreadsheets" allowed anyone to manage a budget or calculate a return on assets. But there was much, much more to be done, including allowing all those new computers to connect together for the exchange of data.

Computers in 1962, and for several years thereafter, were a little like human "savants"; they were able to do calculations at mind-bending speeds, but had no practical sense. Every single step in the process of performing useful work had to be spelled out for the machine by one or more of its acolytes. A sample: "Load Register A with the first number following a semicolon and Load Register B with the second number; then multiply A by B, and store the result in the Cumulative Register." There were abbreviations for the repetitive commands, and the overall vocabulary was called "Assembler Language". Interestingly, these commands had to be translated into yet another language, consisting entirely of numbers, for actual work to be done. In order to join the computer priesthood, one had to learn, to greater or lesser degrees, about these languages.

Access to the wonderful world of computers, I realized, would require one of two things: Assignment to a computer center, where I could learn on-the-job, or additional education. I learned that both approaches were possible in the navy; in particular, postgraduate school was an option for a junior officer with a history of reasonable academic success.

At the same time (1966) a message to all members of the navy (an ALNAV message) advised that opportunities existed in Vietnam, and that volunteers could essentially guarantee their follow-on assignment. Since I was ready to volunteer anyway, this played perfectly into my plans. I was soon slated for a year with the US Military Assistance Command, Vietnam (USMACV) as an advisor to the Vietnamese navy, with a follow-on fifteen month tour at the U S Naval Postgraduate School, Monterey California. My study area would be Computer Systems Management – a cross between a Business Management and a Computer Science curriculum.

These studies were no cinch for me. Some classmates had prior experience with computer systems in the navy's materiel management commands. They seemed to have a significant advantage, as they understood some basic concepts I lacked. But I hung in, and began to enjoy many of the classes. By the end of my program, I regretted not signing up for a Computer Science curriculum, as those were the classes I enjoyed most.

In December, 1970, I graduated from postgraduate school. The navy in its wisdom decided I had been away from sea long enough, and assigned me to two consecutive sea tours. During these years, my closest approach to a computer was the purchase (for $112) of an electronic calculator which could add, subtract, multiply, and divide. I was dazzled – this was such a step beyond the familiar slide rule! Although I still could not imagine the personal computer to come, the calculator surely set the stage for eventual obsession.

In 1975 I finally received orders to a computer-related job. I approached my new assignment with some trepidation – how much would have changed in five years? This was, after all, the fastest-changing field I could imagine.

Upon reporting at the Defense Communications Agency (DCA), I found that I would be responsible for developing a prototype intercomputer network. The computers would be located at the headquarters of the most senior commanders in the armed forces - the Pentagon of course - plus commanders of the Atlantic, European, and Pacific regions. My job was not to develop the computer programs, however, but to oversee contractor teams doing the work, and to identify the functions to be served by the new network. The computer world was indeed changing, but our office was at the center of the change process. This was, with no exaggeration, the exact job I might have imagined, if I were able, at that first meeting with a real computer more than ten years earlier.

In later years, I occasionally told friends that I was inventing the Internet. More accurately, "my" contractor team took the Arpanet, an academic experiment which linked about 20 colleges and research centers, and installed it in a group of military computers. Tasks included very low-level fiddling with packet switching, security management, and distributed database management, as well as the higher concerns of the man-machine interface and the identification of applications which could benefit from the new capabilities. This was a massive effort, spanning many more years than my assignment. When I arrived, the computers had already been chosen and installed at the commanders' HQ's. They were being used for isolated data tracking. Any data concerning another command had to be forwarded by military message, then re-entered in the local system. With some effort, force levels and status were being tracked, and weapons inventories were maintained. But we knew it would be possible, in the future, to make these data and many others, much more up-to-date and accessible around the world.

After three years at the DCA, I was assigned to the Naval War College, in Newport. This one-year assignment was an intense but enjoyable academic experience. Part of it involved war games, as supported by a computer system known as the Warfare Analysis and Research System (WARS). I had found my new home! I made this my first choice for my next assignment (despite a nasty comment from my assignment officer, who suggested I was "retiring on active duty"). My tour of duty on the NWC staff, as Gaming Systems Officer, began in 1979. Once again, I was overseeing a team of contractors, this time developing the replacement for the obsolescent WARS.

By 1979, the development of personal computing was underway. For three years I resisted, while avidly tracking the rapid development of the industry. In 1979, almost all personal computers still relied upon cassette tapes for storage and input/output. Televisions were used for the display. Applications were few, and varied from vendor to vendor.

Osborne 1 Personal Computer

In 1981, the "Osborne 1" was placed on the market, changing the game significantly. This was an integrated system, with two floppy disk drives (each had a 90 kilobyte capacity), and a five inch monitor (displaying 16 lines of 52 characters each – but the user could scroll it left-right to see a full 80 character line). The keyboard latched to the computer case, to form a nicely transportable package. Most importantly, it came loaded with useful software – including word processing and spreadsheet applications. This was the stuff of dreams! I was not alone in my new obsession. At least three co-workers shared my computer fever. We sometimes drove during the lunch hour to North Dartmouth, some twenty miles away, just to see and touch the computers offered at the computer store located there.

In 1982 another key factor fell into place – an excuse! With retirement from naval service coming up in mid-1983, I needed to begin a job search. (At 43 I would not be ready to retire, and besides that, the navy pension was not so generous that a family could live on it.) So, in 1982, I purchased an Osborne 1, along with an electronic typewriter which could be driven from the Osborne. My professional-looking resume may have had something to do with landing a very nice post-retirement position.

In the succeeding twenty-plus years, I went through a lot

more PC's, printers, and programs. Today my story is not at all unusual, and most Americans have more than one computer in their home, and at least one at work. (We also have multiple computers in our cars, refrigerators, and microwave ovens, but that's another story.)

Who could have anticipated the wild ride I was embarking on, when first I saw that computer in 1962?

21

Me and Al Gore

Inventing the Internet

In early 1975, during my final year as Chief Engineer of USS *PONCE* (LPD-15), I received orders to my next assignment – to the "Joint Technical Support Activity" (JTSA), with an address in Reston, Virginia. I soon heard that I would be working on a Prototype Intercomputer Network, in support of the National Military Command Center (NMCC). WOW! This sounded like exciting stuff! I soon ascertained that no one on PONCE could tell me anything at all about this duty assignment. My visit to the Executive Officer was typical:

"XO, I received orders"

"So I hear – you know, it's a payback tour for PG School."

"I guess – but I really don't know how I'll fit in. What's an Intercomputer Network?" What's an NMCC?"

"I have no idea what an Intercomputer Network is - but I do know what the National Military Command Center is."

"And…?"

"Didn't you see "Doctor Strangelove? It's the command center inside the Pentagon where they will fight World War III."

In September, I reported for duty. The JTSA Reston address was an office park – a dozen two-story brick buildings, each with a building number, but no signage announcing identity or purpose. Spooky! The sense of secrecy, however, was somewhat dispelled by the fact that there were no security guards in "my" building. Everyone did, however, have a picture badge, and we were instructed to wear it in a visible location at all times. I was introduced to two levels of bosses, and some peers. Our Branch Chief was Dr. Irving Bialick – a small, pipe smoking Jewish man, with a burning intellect, but a kindly demeanor. His doctorate was, surprisingly, as a Psychologist, not a computer scientist. Nevertheless, it was soon evident that he had plenty of computer systems expertise, as well as a masterful knowledge of, and skill in, inter-office politics.

My immediate supervisor was Dr. Byron B. Moore; he did not care for the name Byron, and was to be just "B" Moore for as long as I knew him. He was a computer scientist with a specialty area of database management systems. He also had extensive programming experience. Monte Welch worked in network systems – improving the availability and reliability of intercomputer linkages.

I also met civilian specialists in computer security and several other disciplines. There were no other Naval officers, but there were several Air Force Majors, an Air Force Lt. Colonel, and an Army Colonel. The Chief Scientist, Mars Champagne, seemed to have more clout than any of the military officers.

I had the good or bad fortune to report for work on the day of the weekly Branch status conference. Dr. Bialick presided, and went around the room asking for status of each of the projects being overseen by the branch. My head was spinning before the meeting was half over. I had fallen into a bowl of alphabet soup! I even began to wonder if these people had a solid grasp of the English language. Our Branch was J310. Holy writ came from the Rainbow Route, and consisted of Buff's, Flimsies, Greens, Purples, and Red Stripes. Important entities were WWMCCS (pronounced "Wimmix"), NMCSSC, WASSO, AFDSC, AFDSDC, the WAD, and the PMO. There were three different meanings for CSC! Technologies included GCOS, GRTS,

WWDMS, IMP's, TIP's and the DN355. I spent most of this meeting just writing down acronyms. A day or so later, I created a cartoon for my own amusement, and that of some later arrivals. Charlie Brown and I had plenty in common.

My preparation for this brave new world was an eighteen month Master's degree program at the Naval Postgraduate School in 1969-70. The degree was in Computer Systems Management – about half computer science, and half management. We were introduced to the PL1 programming language, which ran on an IBM 360 model 67 computer system. We wrote programs, punched them into cards, and submitted them (along with test data, also on cards) across a countertop. 12 or 24 hours later, we came back and picked up our cards and the printout from our run. The printouts NEVER reported a successful run on the first try – at least not in my experience, nor that of anyone else I knew.

Confusion Reigns
(a cartoon created by the author)

We were also introduced to several other computer languages, including FORTRAN, BASIC, LISP, and SNOBOL. We were introduced to the concept of data bases, and database management systems, but never did we consider live interaction between computer and end user. As for inter-computer access, well, one could easily mail a tape from one facility to another!

So now I was introduced to the project. The Prototype WWMCCS Intercomputer Network (PWIN for short, using a compound acronym) was to permit the highest level of U.S. Military Commands to work together. The World Wide Military Command and Control System (WWMCCS) included the Commander in Chief, Atlantic (CINCLANT), CINCPAC, and USCINCEUR, not to mention the National Military Command Center (NMCC) at the Pentagon. Commanders one level lower included CINCLANTFLT and CINCPACFLT, US Army Europe, Naval Forces Europe, and Air Forces Europe. A number of logistic commanders would also be connected. The nation's forces could be more responsive and better prepared, thanks to this program. All we had to do was invent a secure Internet, a way to engage with it, and the tools to make it useful – like a "distributed data base" which could look like a single database, even though it resided in several physical locations, and was continually updated by people scattered around the earth.

One little catch – not only was there no such thing as a secure Internet, there wasn't even an Internet! Its predecessor, the ARPANET, was first conceptualized in 1967 – just eight years before I reported for duty at JTSA. By 1975 an experimental network existed, with about 50 IMP's (Interface Message Processors) nationwide – most of them at university computer science research centers. Recognizing this fact, contracts were established with academics from these centers. A team was already in place, working for us, when I reported for duty.

The development team was led by a young Computer Science PhD working as a post-doctoral fellow at the Center for Advanced Computation, University of Illinois at Urbana-Champagne. A slender, bearded, intense young man, Daniel Alsberg was about my age (36 at the time). He excelled at preparing written reports which encapsulated the research his team was conducting. His major effort, up to this time, had been development of a distributed database management system. Five volumes of dark blue booklets had been produced. Each was a tutorial, as well as a record, of the definition of requirements, function, and design. These books described the division of records into individual 256 bit packets, which could be shipped over a network in any order, then re-assembled into meaningful data records. Each record was identified with the origin, destination, and a small quantity of data, or necessary control fields.

Daniel's team consisted of undergraduates and doctoral candidates. I had less contact with them, but four would be fully engaged in the development over the time I followed the program.

The new investigation, just beginning and my primary focus, was to improve the man-machine interface to the data contained in the computer systems. Up to this point, in the batch processing world, computers gave up their information only grudgingly. A decision maker would tell a technician what he wanted to know, and the technician would prepare a detailed query for the data, punch up cards for submission, and, eventually, the decision maker would have a fat printout of data containing, if he was lucky, the information he actually needed. It was not surprising to learn that the favored information recovery procedure was to pick up a radio handset and ask someone at a lower level of command. Computers seemed to exist to collect information – not to disseminate it.

We hoped to use the power of computing to improve accessibility to computers. To this end, the team proposed to create what I later recognized as a Personal Computer with touch screen interface. They used a Digital Equipment Corporation LSI-11 single board computer in a small wheeled cart. There was a plasma display screen (flat panel, but monochrome – orange on black), with crossing infrared beams to make it "touch sensitive". There were also keyboard and floppy drives (8 inch, of course), and even a hard drive.

This little roll-around computer always created a sensation when it was demonstrated. It was designed to be connected, via cable, to a phone line, whence it could be connected via Arpanet to a site with an interesting database. Obviously we couldn't be using a classified database, such as one showing locations and condition of all Navy ships. These computer geeks, after all, didn't look in the remotest military with their long hair, beards, and "casual" clothing. So the one database we selected for demonstration purposes was an Illinois agricultural database. Pick a county, find out how many cows it has, how much acreage under cultivation for wheat – that sort of thing. This database was a nice one for our purposes, as it had a few graphical features – such as a county map of the state. Using the touch-screen feature, your finger could select a particular county for access to data about crops.

After months of work, which included obtaining and assembling all the parts for the roll-around computer, and significant programming just to get all the pieces to work together, then to interface to the demo database, we were ready to go to the Pentagon for our demonstration.

The demonstration was set for one morning in the summer of 1976. I met the team at a delivery door for the Pentagon. The team, as always, looked like a bunch of bums, even though they may have been wearing their best outfits (it was hard to tell). I had long come to terms with that, and hardly noticed. But I elected to wear my uniform – spotless Tropical Whites with short sleeved shirt, Lieutenant Commander stripes on my shoulders, and a reasonably impressive array of ribbons reflecting 12 years' service including a year in Vietnam. The Pentagon conference room assigned for our Demonstration was fairly crowded with military officers, also in uniform. These individuals were fairly senior – Lieutenant Colonels, Colonels, and one- or two-star Generals, and their Naval equivalents.

"Why are you dressed like that?" asked Daniel. Service members in the Washington area in those days wore uniforms just once a week, as a rule, and I had seldom if ever worn the uniform while working with our development team.

"I don't want to be confused with you people", I joked. I could see immediately that was the wrong thing to say. Daniel clearly didn't find it funny, and I later decided he had passed my remark along to the other members of the team.

Out demonstration went well enough, although our Arpanet connection failed early in the demo. That's where the hard disk showed its purpose – a previously loaded copy of the database was available to continue the demonstration, even though we were not now connected to the Illinois database. It appeared to me that none of the attendees even realized their data wasn't live. (I did hear a few caustic remarks, however, about just how useful an agricultural database was to fighting a war.)

I never apologized to the development team, to my regret. The project was over, the money used up, and I went on to other projects within DCA. I had no further contact, professional or other, with the members of that team, and they faded from memory. I cannot, today, name any one of them – even "Daniel" is a guess – but I would not be surprised to learn that one or more of those kids became important contributors to the development of the computer-based life we live today.

22

Passage to Djibouti

IN 1963 I was an Ensign on the destroyer USS Bigelow (DD 942). In November, I had the opportunity to trade places for a week with an Ensign assigned to the submarine USS *Sirago* (SS 485). Our two vessels were ordered to proceed, each independently, from Karachi, Pakistan to Djibouti, French Somaliland. Distance was approximately 1,600 nautical miles; an average speed of 12 knots would be required over five and a half days.

I was Bigelow's Assistant Communications Officer (another Ensign, with a year more experience than I had, was the real Comm Officer), and therefore I was eminently available for unnecessary assignments and adventures. Three months earlier, I served as the practice dummy in a highline transfer between ships. Also during the deployment, I took an even more exciting ride, being hoisted from Bigelow's deck in a "horse collar" suspended by wire from a helicopter hovering above our deck. The command didn't need to worry about me - I was the supernumerary.

Bigelow was perhaps unusual in having so many Ensigns. Once as the entire officer complement straggled down the pier to a Destroyer Tender for a repair conference, I realized there were eleven of us Ensigns! The ship's skipper was a Commander, the Executive Officer was a Lieutenant Commander, and there were four Department Head LTJG's; otherwise every officer was an Ensign. This was a good ship on which to gain experience - which is obtained, of course, from making mistakes!

This particular adventure began on the morning we were to depart Karachi, Pakistan after a five day port visit.

We were part of (virtually the entirety of) a Middle East Force, responsible for showing the flag in that part of the world. At the time, this force consisted of our destroyer, one submarine, and HMS *Meon*, a British seaplane tender at Bahrain. From the Mediterranean, we transited the Suez Canal and then the Red Sea, then conducted visits to Ethiopia (Masawa), Aden (a British colony then, part of Yemen now), Saudi Arabia (Ras Tanura), Kuwait (Mena-al-Ahmadi), Bahrain, and Pakistan (Karachi).

Karachi was a city alien in its language, garb, smells. Camels were common sights, and so were horse-drawn wagons. Dress included long, flowing garments and western working garb - not many women were in view, but none wore the all-enveloping hijab so popular nowadays. Craftsmen by the hundreds toiled and tended their stores, offering wooden bowls, platters, chess boards, and boxes, inlaid with brass and ivory (or perhaps bone). Camel saddles, usable as a hassock or stool, were a popular memento until vermin began emerging, weeks later, and quite a number of them made at least part of our voyage home in staterooms and equipment lockers. A shoemaker came aboard on the day of our arrival, drew an outline of a customer's foot, and delivered a pair of camel leather shoes before we departed. As we prepared to depart on a Saturday, I reflected that this had been quite the unique port visit.

Friday before our departure was November 22, 1963. President Kennedy was shot at 12:30 PM, Dallas time, 10:30 PM in Karachi. Word of his death reached Bigelow sometime during the night, by means of the Fleet Broadcast. It was announced on the ship's announcing system at 7:00 A.M. Saturday - the day of our planned departure. I was awakened by the announcement that, halfway around the world, the President had been shot and killed. Although shocked, I also felt quite detached, because of the distance. In any event, I had no time to stay and discuss - I was off to join USS *Sirago* (SS 485) for transit to that next port visit. Walking down the pier, I carried my gear in a canvas sea bag. Not much - just clothing for five days; working khakis only - no dress uniform would be required.

Sirago was a diesel boat, designed during WW II, built in 1945. She displaced 1,570 tons, was 312 feet long, and capable of 20 knots surfaced, 8 kts submerged (for short periods). Her test depth was 400 ft, and her complement was 10 officers, 71 enlisted. There were four diesel engines and ten torpedo tubes, with a maximum loadout of 28 torpedoes. She was equipped with a snorkel, to permit operating diesel engines while submerged to periscope depth.

On deck to board the USS *Sirago* submarine

My first impression upon entering the vessel: Cramped, in spades. We entered the submarine via the conning tower, where there was a watertight door leading to a ladder up to watch stations, and down (through a watertight hatch) into the pressure hull and the Control Room. Below that was a main fore and aft passageway. Every 20 feet or so, this passageway was interrupted by watertight doors, complete with shin-buster coamings for secure sealing. A small crew's mess. and a tiny wardroom mess (for officers) provided a place to eat, do paperwork or read. Bunk rooms were divided by the main passageway. Officer sleeping quarters were a little more private, in tiny staterooms. I shared the XO's stateroom; his bunk was at a convenient, three-foot height. Mine was directly on the deck, and I had to crawl into it, rounding a corner where his desk and wardrobe abutted. The entire room was about six by eight feet, including the bunks, a folding sink, and a combination wardrobe and desk for the XO.

Claustrophobia would be disqualifying for a submarine assignment, even a temporary assignment. Where I felt any such feeling, I consciously concentrated on "cozy", not "close". Walking aft, one passed through the engine room, with those four big diesels. Farther along, there was a crew bunkroom, then the after torpedo room, where four tubes pass through the watertight envelope. Going forward from the control room, you quickly reached the forward torpedo room. Those big 21 inch torpedoes were very much in evidence. So were sailors bunks - many of which were directly above or below a torpedo. There were more crew members than bunks, so some of them were shared in a "hot bunking" arrangement.

Everywhere there were wires, pipes, and valves. When not on watch, the sailors spent most of their time learning new systems, or teaching those systems to the new men. Ideally, every crewman knows how to operate every system; this could be the difference in floating or sinking in combat.

One system everyone had to master (even a visiting Ensign) was the sanitary drains. Using a toilet involved choosing, opening, and closing valves in the proper order. Probably there was no chance of sinking the boat by making a mistake, but the operator could be soaked with a most unsavory spray. I quickly mastered that system. A submarine sink was hinged at the back; lower it to wash, raise and close it to empty it (there was no drain in the basin itself). Faucets were spring-loaded levers, to avoid wasting water.

There was very little room allocated to general storage. Luggage, however, could be conveniently stored in showers - those were not usable at sea for their nominal function. There was simply no water for such indulgent use.

Watch standing in the conning tower was a treat. Fresh air, daylight or starlight, was otherwise non-existent. Most of our transit was on the surface. Cruising at 12-16 knots on our southwest course, we paralleled the coast of the Arabian Peninsula (present-day Oman and Yemen). Weather was warm and humid, as one might expect. In my brief experience, there was no stormy weather, but I have no doubt the submarine could submerge to escape the worst effects of any storm.

As you may well imagine, submerging on purpose does not come easily to a surface sailor. Nevertheless, as a matter of continuous training, Sirago made numerous dives, even though the more efficient mode of travel was on the surface. Each was announced by several "AOOOGA's" on the klaxon, followed by the announcement: DIVE, DIVE, DIVE. Watch standers on the conning tower plunged immediately down the hatch into the interior, via the single vertical ladder. Last man down was the Officer of the Deck, unless the Captain was there - then he would be the last man down. As he entered the control room immediately below, he would call "LAST MAN ", and watch standers there would slam the hatch and spin the watertight dogging wheel to seal the boat. With constant practice, this evacuation could be accomplished in about 30 seconds; in an emergency, diving would commence even before the bridge was fully cleared.

Submerged travel was an altogether different thing than steaming on the surface. Sudden quiet replaced the continuous thrum of four diesel engines. A slight overpressure made everyone clear their ears. Now we were dependent on our main batteries to power the propellers, and speed was limited to an absolute maximum of 8 knots; more typical was 3-4 knots.

Although we were essentially undetectable while running submerged, there were limits to how long we could run. This sub was able to use a snorkel to run engines and re-charge batteries without surfacing. Once a day, to maintain proficiency, the snorkel was used in this fashion. If the surface was rough, waves could wash over the snorkel, forcing the engines to obtain their oxygen from the interior of the submarine - this was VERY noticeable to the ears, and the Captain made sure this happened. Once again, we had to adjust to the pressure change.

After a night of surface travel in those waters, Sirago would invariably have a collection of flying fish on the deck grating. In daylight hours, it was fascinating to me to see them "flying" along, propelled by a sweeping motion of the tails on the water surface. Their large, wing-like fins then took over, and they could glide hundreds of feet through the air. Clearly, however, their night vision wasn't great!

Another visitor came our way during the transit: a green parrot, with no land in sight and obviously in need of somewhere, anywhere, to rest, alit on the conning tower, practically in the midst of the watch standers. He was easily captured, and one of the sailors volunteered to take the bird below for some food. Predictably, the bird didn't like the atmosphere below decks, with no sunlight and no fresh air. Panicked, he flew wildly into one or more obstacles; he was buried at sea.

The wardroom wasn't just for eating, it was also a miniature recreation hall for the officers. Movies were carried on all ships in those (pre-video) days. They were bulky, on 16mm film, and the projector, of course, made a lot of noise. Movies were exchanged between vessels whenever possible, and the projector noise was accepted for a few hours of distraction. On Sunday of our transit, the movie "Judgment at Nuremberg" (a six hour extravaganza) was shown. I don't think any single individual watched the whole thing; watches continued, officers sat in for a time, then moved on. I was told that I could skip my bridge watch to stay, but didn't want to miss my time on the conning tower.

On the fifth day of our transit, we entered Djibouti. It was located in French Somaliland, which was founded around 1887, overrun by Italy during WW II, and restored as an overseas territory of France in 1946. In 1967, French Somaliland would be renamed the "French Territory of the Afars and the Issas". In 1963, I saw nothing justifying a visit - let alone owning the city. Evidently the French reached a similar conclusion; in 1997, the territory became the independent country of Djibouti. Sirago moored to the same concrete wharf as Bigelow, and I was very pleased to re-join my ship, and some semblance of civilization. The first thing I did was shower and change. After that, I tried to catch up on the state of our nation following the assassination, but with no TV, and with news magazines lagging by two to three weeks, none of us participated in the national horror show involving Lee Harvey Oswald, Jack Ruby, the state funeral, and countless conspiracy theorists.

Still, my "Passage" was a memorable experience, and I was happy to rejoin a more nearly normal life.

23

Risks And Rewards

There haven't been too many risks in my life (we will disregard two tours in Vietnam for the moment); mostly, it's been just one foot in front of the other. But as I chose a naval career before I was out of my teens, there was always the inevitable prospect of leaving that chosen life - someday. Even so, taking the actual step felt risky.

Departure did not consist simply of writing a retirement letter. If fact, it began four years earlier, when I was just 39 years of age, with 16 years of service (20 years being the minimum to qualify for a retirement pension). I was finishing my year as a student at the Naval War College, and submitting my preference card for the next tour of duty. I would have liked to go to an assignment as commanding officer of a ship, but realized that, for reasons I won't go into, that was not going to happen.

I became aware that a position was opening in the War College staff as the Gaming Systems Officer at the Center for War Gaming. I was qualified for this opening, by virtue of my master's degree in Computer Systems Management. I also had served one tour in that sub-specialty at the Defense Communications Agency's Command and Control Technical Center in Reston, Virginia.

To be assigned to the War College job, all I needed to do was submit my preference card requesting it, while also requesting it internally by memorandum to the President, Naval War

College. I knew they would not find a better qualified candidate, and I could do it without relocating.

All I had to do was give up any hope of future sea duty assignments. I loved sea duty. The sights, smells, and even the rolling and lurching of the ship in open seas were deeply rewarding. I had never seen an ocean when I entered the Naval Academy, and found myself absolutely entranced by the first experience, a year later on midshipman cruise. I was assigned to a Gearing class destroyer, commissioned at the end of WW II. I could stand in a sheltered spot on the main deck, eight feet above the ocean surface, and stare into the spreading bow wave for hours on end. The change in sea color from gray, to green, to blue as we progressed farther and farther south seemed almost magic. From the bridge, I could watch the bow plunging into rough seas, and see the spray blasting its way even higher than the bridge. These experiences would become strictly history, no longer possible future moments, as soon as I asked for more shore duty.

Well, I thought, there aren't many ships left with that low freeboard, anyway. As a Commander (never to be confused with commanding officer), I was offered a job as Materiel Officer on a major staff. I would be responsible for tracking readiness and repairs, arranging fuel stocks and spare parts. An office job, even if located on a ship. And it would be a very large ship - no communing with the sea gods there.

Furthermore, shore duty would allow me to be an at-home husband and father - another job I loved. My wife and I had children four, eight, and nine years of age. The eldest had Down syndrome, and was enrolled in "special ed". During my year as a student, he had made surprising progress, learning to read and write in Newport's Mumford School.

Finally, I felt that my job in War Gaming would be technically engrossing. In addition to overseeing the development of a new gaming support system, I would work as an umpire during war games.

So, I filled out my preference card to indicate my first choice was the job at the Naval War College. I also indicated I would be submitting an internal request to the NWC President. In effect, I was telling the Bureau of Naval Personnel "Don't even bother to look at my second choice." This had a predictable but unsettling result: an evening call from my "Detailer", who accused me of wanting to "retire on active duty." This call confirmed my earlier conclusion that this request would write finis to my naval career.

In due course, I received the assignment I had dictated. Then I began to wonder if there was a new hazard - that somehow the Bureau would see to it that I was forced out prior to qualifying for a pension. I needed four more years, but my assignment to the new job was for three years.

Ultimately, of course, that Detailer moved on, my assignment was lengthened to four years, and I was able to retire from active duty at the end of that time. Better yet, an addition to the family coincided: my wife bore a son on the very day of my retirement from active duty, and I have frequently referred to him in the 30+ years since as my "retirement bonus". On top of everything else, my experience with development of a war gaming system led to civilian employment after military retirement. I worked in that field for eleven years, thus compiling a different kind of "retirement bonus."

24

SELECTED

IT SEEMS I always had an academic leaning. Before I was old enough for school - perhaps the year I turned six - I enrolled my three-year old baby sister in games of "school". We would open copies of Readers' Digest and pretend to read, then we would have writing lessons, scribbling on pieces of blank paper. Both subjects were seriously handicapped because neither of us knew how to read or write.

There was no such thing as pre-school, or even kindergarten, in Tennessee in those days. But once I turned six, the games began. I was somewhat disillusioned on my first day of school, however. Sister Mary Rosario assigned us homework - we had to copy the entire alphabet! To my mother that afternoon, I lamented having twelve years to go. Little did I know.

Fast forward through NINETEEN years of school. In some of those years, I was a Teacher's Pet, because I was such an attentive, smart student. (That way I could slack off on the detested homework, and still get by in the classroom.) Anyway, at age 36, I was thirteen years deep in a navy career, on assignment to the Defense Communications Agency's Command and Control Technical Center in Reston Virginia. The year was 1976, and my assignment involved development of an inter-computer network to serve the most senior commanders of U.S. Armed Forces around the world. I was deeply immersed in an arcane field which was, in retrospect, developing a military version of the Internet - which didn't yet exist. I found it endlessly exciting and challenging, and I loved it. Beyond the packet switched movement of data, we also worked on distributed data management, and developing an improved Man-Machine Interface - personified in a "smart terminal" which was a lot like a personal computer connected to one of those networked main frame computers.

For better or for worse, in the navy nothing lasts forever. I was soon forced to consider my next assignment.

"Mr. Heins: mail for you", said Doctor Bialick. I had encountered him in the hallway of our basement office complex. Or maybe he had encountered me. He was holding a single piece of mail.

This was unusual on many levels. Doctor Bialick, my Branch Chief, was a little above my daily circle. He was my boss's boss, we seldom had direct conversations, and he NEVER delivered my mail. A small, wiry, balding man, he had the look of a serious runner - at least until you remembered his ever-present pipe. He was a "Doctor" of psychology, not medicine. Early on, I had been surprised to learn how many psychologists it took to develop a computer industry. Also unusual because the organization seemed to favor first names. About half the staff was civilian, the other mid-grade officers from the Army, Navy, and Air Force. Everyone used first names.

The letter was clearly official, with a return address indicating the Bureau of Naval Personnel. Dr. Bialick waited for me to open it, evidently believing it was for both of us. I opened the envelope carefully, and extracted a crisply folded single page.

The letter was a little less terse than some; fewer acronyms, fewer OPNAV gobbledygook phrases. "You have been selected to attend the Naval War College senior course, class commencing 1 September 1978," it said. It went on to explain that I had thirty days in which to respond, using the enclosed forms to accept or decline this "nomination".

"Well? What's the news?" puffed Dr. Bialick.

"I've been approved for the Naval War College," I explained; "Looks as though I'll be leaving you next summer." But I knew I would be checking my options, as I was not at all sure I wanted to go to Newport, Rhode Island. I had spent a dark, cold spring there at Destroyer School, ten years earlier.

"Well, congratulations!" said he. There was no indication he would have any trouble replacing me in his organization, no matter how essential I felt.

After reflection, I decided a twentieth year of education was the best offer I would get. I reported to Newport in August, 1978. Along with wife and three children, I was assigned student housing at Fort Adams, a relic of the War of 1812. Our residence was a town house in a complex arranged as two facing rows, with a nice brick patio or courtyard between the two rows. All residents were in the same assignment, and many friendships were built. It was one of the most satisfactory arrangements of my life; and it was almost certainly the most enjoyable residence of our married life to date for my wife. For starters, I was to be home the entire year, rather than deployed on a ship. Even though I would have class and study assignments, I was available to help out at home.

In the second trimester of the program, we were exposed to several war games, using the Naval War College Center for War Gaming. The building contained some forty "command cells", intended to replicate individual commands afloat or ashore. Participants were bombarded by realistic sounds (the constant rattle of a teletype machine, and the sometimes cacophonous demands of multiple radio-telephone speakers). These communications connected the commander to other decision makers and subordinate units. Each of us students had spent time in just such command centers, and were familiar with the strain of deciding whom to listen to. Charts were maintained showing positions of friendly forces, and intelligence estimates of opposition forces. Participants had to decide what steps their own forces would make. At the end of the exercise, all could gather in an auditorium, where two giant screens (25 feet high) could display the actual tracks of simulated forces.

Shortly, I learned more about those displays. They were controlled by a very ancient computer - originally used on a ship in the 1960's to control Navy Tactical Data System (NTDS) displays. This computer could handle a total of 48 tracks, and its processor was so old it used 386,000 tiny iron rings in the memory to hold the bits ("1's and 0's") computers require. (Each ring was threaded by three wires to control and sense a magnetic field.) More modern computers used transistors to represent those 1's and 0's; even in 1978 a typical memory size registered in the millions of bits. Clearly the computer needed replacement, and acquisition was underway.

I was in the right place at the right time. The War College needed an officer with exactly my background to represent it in acquiring the new gaming system. I submitted my request for assignment at the end of my tour as a student. BuPers hates it when you "write your own ticket", so my detailer accused me of retiring on active duty - but he did write the requested orders. In the fall of 1979 I transferred from student to staff of the college. That assignment took me to retirement in 1983. By then I was "expert" in computerized naval war gaming, and joined a local company which had a contract to assist the War College in effectively using the by-now delivered system.

That system showed significant weaknesses, however, necessitating an Enhanced gaming system. After a year away from the acquisition process, I was hired by the developer, Computer Sciences Corporation. Now I was responsible for producing the new version. Eventually I stayed with that company for ten years - still in Newport, despite its dark and cold Springs. At the end of that period, I was forced to choose between company/job and location. After 15 years of residence in Rhode Island, though, it had become home. I left CSC, and eventually drifted into real estate, where I labored for another 15 years before REALLY retiring.

Looking back, I sometimes think, "How did I get here?" One answer is that crisply folded letter in 1976, announcing that I had been "SELECTED".

25

The Risk of a Lifetime

Risk in life is a little like greatness. According to Shakespeare, "some are born great, some achieve greatness, and some have greatness thrust upon them". Everyone faces risks in life; some are freely chosen, some are unavoidable. Some risks are so overwhelming, we don't even think of them in terms of risks. Rather, we simply pity the poor child born into impossible conditions.

My life, on the other hand, has been marked by few risks. When I did make a life decision, it was based on avoiding the greater of evils, or on a clear preponderance of benefits. Is it different for anyone? Every choice is made, inevitably, after a conscious or unconscious mental inventory of benefits and risks. You might not agree with the weighting used by someone else – is excitement really that important? – but it's always possible to hypothesize logic in a choice.

That being said, some decisions involve much greater risks and rewards than do others. In my experience, leaving a ten year employment with Computer Sciences Corporation (CSC) was my riskiest and most difficult decision to date. For the last nine months with CSC, I had been promoted to a position in New Jersey; I was commuting on a weekly basis to work there. This decision had the effect of abandoning a rewarding career at age 54, with no real idea of what I would do next.

As hinted, there was an element of choosing the lesser evil.

By the time I told CSC I would not move to New Jersey to continue employment, I had been living in Rhode Island for 16 years – second in duration only to my 19 years in Nashville, before leaving home to attend the Naval Academy. I had come to feel at-home in Rhode Island, with ties to church, schools, and friends. Moving to New Jersey would mean starting over socially, and would impact family – especially the one child still attending school at the time.

A degree of job burnout contributed to my decision. Increasingly, I felt a tension between product development schedule and product performance issues. Much of my job, it seemed, was making excuses to the government customer for why we didn't meet schedules (always overly optimistic) and product performance goals (for some reason, perfection continued to elude.) I found the idea of stepping off the treadmill and relaxing at home very appealing.

Finally, there was one specific benefit I hoped to protect: my son Michael, who was then a 24 year old young man with Down syndrome, was associated with Middletown's James L. Maher Center, which was dedicated to providing an appropriately challenging and protective work environment for Michael and his peers. In addition, Mike had a rich social life outside the work center, including studies, dances, and physical exercise programs. I had researched his prospects in New Jersey, and found nothing comparable. Rhode Island was simply years ahead of any other state, so far as I could determine, in providing a rich and rewarding life for my son.

The potential down side for my decision, so far as I could see, boiled down to money. A small navy pension assured me that I could pay the mortgage, but I had no specific work prospects, and I began to realize my work experience and skill set did not automatically translate to other jobs. As the decision date approached, I found it more and more scary. I wavered, telling myself "I've moved plenty of times – this wouldn't be so bad."

I found myself tossing sleeplessly at night. Sometimes I wanted to continue as I was – essentially on permanent travel status, staying during the week in a New Jersey apartment, traveling to and from Rhode Island Friday afternoon and Sunday evening by car, train, or plane. But the government monitor became increasingly nasty about the expense, and some of my CSC management were restless with my "short work week" (ignoring the fact that I worked more than 60 hours exclusive of travel time). Eventually, I formally notified my manager that I would not be accepting relocation to New Jersey. In return, I was given a layoff letter, and ended my weekly travels.

The scariest part of this decision was recognizing that despite my young age (54), this could well signal the last time I ever held a working position of significance. For many years I had been, to someone, "the boss". This was true for my final ten years on active duty, as well as in my ten years with CSC. While at CSC, the pay and the responsibilities were far above average. I knew it would be difficult if not impossible to replace that revenue stream.

Shortly after I began my new life, I realized there was another fairly serious hitch: without employment, there was a serious problem of BOREDOM. Before long, I attempted to solve both boredom and poverty by entering training as a Real Estate agent. After netting some $2,000 in my first year as an agent, I realized I had solved the boredom problem, but not the financial one. I also learned I was poorly suited to the roller coaster world of real estate sales, where a solid agreement could collapse with little warning, so that a sale potentially worth several thousand dollars became just an anecdote.

After three years of struggling with real estate sales, I moved into appraisals, and found some unexpected benefits. For one thing, I was now doing tasks myself, rather than managing technical experts who actually accomplished work. This was in welcome contrast to my years as a computer systems manager.

I found each appraisal report to be like a little puzzle, which I could solve on my own. There were no more paydays of thousands of dollars, but there were far more regular paydays of hundreds. Never again did I approach the $80,000 per year levels of my CSC days, but with the navy pension (especially after Social Security kicked in) I felt at peace with family finances.

As I had hoped, Michael's life continued to flourish. For several years, he worked at the Navy Base in food service. It was hard work, but he was proud to be doing important work which benefited someone. At the same time, his social life continued, as he attended Salve Regina University's Learning Unlimited, swam regularly at the YMCA, took weekly Karate lessons, and served mass at our parish church. Ultimately, he moved into a Group Home, where he now lives with four other friends, and basks in the loving atmosphere.

In making this riskiest of all my life decisions, I can look back and remember with pleasure that there was never any external opposition to my decision. The only votes that were important belonged to the family, and all were completely supportive. The only real opposition to my decision consisted of my own fears, which proved needless.

WINDING DOWN

26

A Lifetime Of CAUTION

My financial life divides semi-neatly into three parts. First there was my childhood, when I was poor but mostly unconcerned. I knew there were rich people - my Dad repaired their furniture, and I sometimes helped him pick up and deliver it - so I saw how the other half lived. Still, everyone I personally knew, including virtually all schoolmates, lived lives similar to my own. Most had nicer cars, but I had no knowledge of other homes, so did not feel disadvantaged to live in a one-bath house. Indeed, I still believe most of them probably had just one bath. At least we had indoor plumbing, unlike my mother's relatives in rural northeast Georgia.

Part Two was my four-year stay where Severn meets the Chesapeake, at the U.S. Naval Academy. Ironically, no soviet communist institute could have arranged a more obsessively leveled playing field, financially speaking. Every midshipman was paid, dressed, fed, educated, shod and shorn identically. Privileges varied strictly accordingly to performance and years of residence. Although I knew certain classmates were much better-off financially, it simply made no difference from day to day. In effect, we were in a four year monastery. Until the end - when it was time to choose wheels for a resumption of life after the monastery. At that point, there was a significant difference in who bought the Corvettes and who bought the VW's. My choice, predictably, was a VW Beetle, for $1,200. I was determined to take on as little debt as possible. After all, I was well aware that the pay of a newly commissioned Ensign would be $222.15 per month (plus a $47.88 "subsistence allowance"), before taxes.

And so it was that I began Part Three of my financial life. At the very beginning of my naval career, I felt it was unfair that certain individuals had much more take-home pay than I did - specifically the married men, who received a handsome $85 per month housing allowance. But then I could live for free onboard the ship, and there was this thing called the snake ranch: by pooling our money, three or four bachelor officers could afford a very nice rental unit while the ship was in home port - in our case on that first ship, four of us lived directly on the seawall waterfront of Atlantic Beach, Florida. Two years later, on my second ship, three of us we rented a waterfront house at Mission Beach, California - a suburb of San Diego. In retrospect, I realize I never had so much spendable money.

Then I got married.

What a difference a wedding ring makes! Even though my pay was rising, and even though I now qualified for a housing allowance, there would never again be an oceanfront residence in my plans. Instead, I could expect to be responsible for the expenses of a family from my income alone. I became a saver and an investor. Several money-making schemes presented themselves, and I'm sure they all made money - for someone else. There was the Limited Partnership which owned an apartment house in San Diego. I tried investing in commodity futures at one point. And then there was the stock market - by far the best of the lot, in my experience. In addition to common stocks, I invested in mutual funds, to my eventual dissatisfaction.

When my spouse and I finally had an assignment likely to last more than 18 months, we bought our first house. It was a great place to live, in Herndon, Virginia - a Washington suburb. Taking out a mortgage for more than four times my annual income seemed frightening, but these were boom times, this was a boom town, and, after all, we did need a place to live. When we sold and moved, after three years, there was a nice profit to plow into those savings (which had NOT grown for the three years). In some respects, this experience was the highlight of my investment life, in return on capital, and formed the basis of future security as I used the profit to buy carefully selected stocks.

In the remainder of my parenting years, My financial outlook was ever cautions. Cars were a succession of Plymouths. Homes were apartments or navy quarters. Eventually we washed up on the shores of Narragansett Bay, and after a single year in quarters while attending the Naval War College, we bought a home in Middletown. Once again, the financial commitment was frightening, but inevitable and ultimately satisfying. Again, we knew contemporaries in far more impressive homes, driving far more impressive cars, but caution limited our choices. Thankfully, my spouse and I were in complete agreement on priorities. We were always ready to invest in our children, but otherwise saving for the future took precedence.

As my twentieth year of naval service approached, I had to think about retirement. The navy pension would commence immediately upon retirement, but would amount to only approximately $20,000 per year. In 1983, that would pay for mortgage plus food or heat, but not both. A year ahead of retirement, I bought a personal computer with attached typewriter to allow me to create a business-like resume. As it turned out, I had a generous offer from a local entrepreneur who won a contract with the Naval War College to provide war gaming support using three experts - retiring naval officers with extensive experience using the newly operational gaming system. That meant two friends and myself. After a year, I moved to join the much larger company – Computer Sciences Corporation (CSC) - developing the next-generation system. Not only did I have postgraduate education in computer systems management, but I had a thorough understanding of the needs to be addressed.

So in the ten years following my navy retirement, I had more income than I ever earned in the navy, PLUS navy retirement pay. Nothing really changed in our life style, but children began soaking up available income for college expenses. Our cars and houses continued to be modest. Retirement savings continued to grow, taking maximum advantage of a 401(K).

All good things come to an end, however. After ten years working for CSC, I had to choose between being laid off and moving to New Jersey. So in 1994, at age 54, I made the momentous choice. I left the computer industry, and signed up to sell houses. While there were some social benefits (I had quickly become bored with my own company), there was precious little financial reward. Moving into appraisals, rather than sales, the income became a little more reliable, and I found the work far more agreeable than sales. This lasted until I reached the magic age of 70. By then, my social security pension had kicked in. At age 70 and a half, I was required to begin withdrawing from my 401(K) (now converted to an IRA Annuity). After a lifetime of cautious spending and saving, I no longer needed to work for a living. Interestingly, I still seem to be compulsively cautious in spending, and still seek to save. The habits of a lifetime leave a mark.

27

AN ALTERNATE LIFE

WHAT IF?... How might my life have been different, if certain decisions had been made differently?

Despite the many years involved, the number of major crossroads are surprisingly few. Science fiction tales suggest that every decision is a fork in the road. Each decision by every person initiates an alternate reality, playing out forever in one of the infinite number of parallel universes. But it seems to me that most alternates are different only in the most trivial details. My choice of spaghetti or lasagna, to watch "Castle" or "Chuck", would have no impact the next day – let alone the next decade.

Still, the idea is intriguing, and I can think of some choices which would have resonated for the rest of my life – in fact, which do so resonate. My most fundamental choice, in lasting impact, was in 1959, when I elected to accept an appointment to the United States Naval Academy rather than remaining at Vanderbilt University.

Since I was in NROTC already, I might have served in the Navy nevertheless, but I would have been commissioned one year earlier, been assigned to a different ship, known a different set of navy friends, lived in different cities. It seems very unlikely that I would have met the same girls at the same time, and so would have lived out my life with a different spouse, and different children.

With deeper roots in Nashville, however, perhaps I would

have returned there after my obligated service. Both my brothers graduated from Vanderbilt as Electrical Engineers, and left for employment with national companies in distant cities. In those days, companies sent recruiters to major universities to hire engineers. "Job competition" had an entirely different ring than it has today. Gordon graduated in 1952, and relocated to Milwaukee to work for Allis Chalmers for the magnificent sum of $500 per month. Jimmy followed along, graduating in 1955, and went to Westinghouse in Lima, Ohio. I don't remember ever knowing his starting salary, but he left home in a brand new, 1955 Ford. Despite their decisions, I felt, at the time, that I would remain in Nashville if I could. I had close friends from school, some of whom went back to day one of first grade. As a matter of fact, I still have some of those friends, despite my own peripatetic life.

It is even possible that, had I remained at Vanderbilt, I would have chosen a different field than "Double E". I knew there was a better chance of working in the mid-south as a civil engineer than as an electrical one. But I did like that uniform! And the study of Naval History, especially the Pacific campaign of the Second World War, was alluring. In high school I had applied for, but not received, an NROTC scholarship to Vanderbilt. But I signed up for the program as a self-supported student, nevertheless. Midway through my freshman year at Vanderbilt, I received a grant-in-aid from the school to finish out the year, an acceptance by NROTC, and an appointment to the Naval Academy. Feeling less in common with Vanderbilt at large than with the NROTC, I believed the academy would be a better preparation for a naval career than Vanderbilt, and so opted to re-start my college experience at Annapolis.

Partly my fascination with the Navy had to do with a film I had seen, "Mister Roberts", where Henry Fonda bravely defied an overbearing and undereducated Commanding Officer. Years later, after a naval career, I realized Mister Roberts was a disloyal malcontent who brought as much unhappiness on the ship's crew as he did upon the Captain – but that's another story.

So… what would have been different? Pretty much everything? Different wife, different children, different home… a different life. Perhaps I never would have entered the computer field, so my entire working life would have been different – but here I have a doubt. I think it is at least possible that I would have had a very similar Navy career, including a tour in Vietnam, Postgraduate School in the same curriculum, and similar though not identical assignments. There is even a reasonable probability I would have come to the Naval War College, and remained for a staff assignment.

More importantly, how would the differences have affected my life satisfaction and happiness? Somewhere along the line, I became convinced that you are as happy as you decide to be. Satisfaction in life depends on what you value, beginning with health, continuing with satisfying work, friends, family, and financial adequacy.

Regarding financial security: I have known a few very wealthy people, and have read about more of them. Clearly, unlimited wealth does not make everyone happy. (Besides that, almost no one feels they have "unlimited wealth" – no matter how it looks to the rest of us.) I am not wealthy, but do have "enough", after a lifetime of mortgage and car payments, and participation in various savings programs.

My health, like my wealth, is adequate. I know I could be healthier through more exercise and less food, but that seems beyond my degree of self mastery. Is my good health due to my choice of schools, back in 1959? Maybe, and maybe not. Active military service may have kept me more active and healthier than civilian life would have – but if I'm right about pursuing a naval career even if I didn't attend the Naval Academy, then that's moot. Furthermore, if I had lived a civilian life, possibly I'd have become connected to a health club or golf course.

Friends and family probably would have been little different. Although spouse and children would have been different people, human nature and moral standards would have been the same. I'd have been drawn to a spouse largely for her inner qualities, would have been faithful to her, and together we would have tried to impart a sense of honor, responsibility, and love to our children. It would have been a delight to know my childhood friends more closely over the years, but if I had followed the naval trail as I suspect, a few close friends would still have been developed and prized.

So, bottom line, different track, same line. A really different life would have required a different person with different values, different skills, different leanings. Sure, if I had gone into bank robbery life would have gone differently – but that life was sealed off for me in early childhood. Maybe Popeye said it as well as it can be said: "I yam what I yam."

28

Making Me

As a certified and aging Introvert, I spent a fair amount of time considering just how I came to be who I am. Schools attended, career choices, early religious instruction, regional culture, all contributed. Many of those must be attributed to my parents - it was they, not I, who decided I would live in Nashville, be a Catholic, attend Catholic schools. It was they, especially my mother, who did their best to endow me with a sense of right and wrong - even before I started to school. And neither of them "spared the rod" with their children, lest we be spoiled.

Most of those lessons on being honest and trustworthy came from my mother. She had a college education (as a nurse), and always encouraged us kids to read, and to do our best in school. Besides that, she was the one who was always there - especially in earliest memories.

My father, on the other hand, had only an eighth-grade education, and although he fully supported his children's attending high school, he saw no reason he should pay for them to go to college. He worked five and a half days a week, so we children saw a great deal less of him, than we did mother.

Daddy's eight years of school did not mean he was in any way a dummy. He had a strong functional grasp of the mathematics he needed to own and run an upholstery business, and to calculate, for example, the materials and amount of cloth or leather he would need for rebuilding and covering a sofa.

My father was born in 1898, and was 42 years old the year I was born. Sometime during his working years before my birth, two significant things happened: An accident cost him an eye, when an upholstery spring snapped and sent a metal fragment flying. For the rest of his life, he wore a glass eye - a very good cosmetic match to its functioning mate - but one result was a number of automotive sudden stops and minor fender benders. Aside from that minor irritant, his being monocular seemed to give him surprisingly few problems. He could thread a needle about as well as anyone else, and he seemed to have no trouble reading a newspaper or magazine, although I don't remember him reading books.

The other significant work-related event took place during the Depression; Daddy was just one of the employees, but a diligent one who had been saving part of his wages. When the owner found himself strapped for cash, my father helped out in exchange for a full partnership. The business was henceforth known as "Meinhardt & Heins, Upholsterers".

Ownership did not purchase a life on easy street, however. It brought more responsibility and stress, probably contributing to two "nervous breakdowns" he suffered - once when I was perhaps four years old, and once when I was 12. These breakdowns were stressful on the rest of the family. I have a vivid memory of being hustled out the back door of our home, to take a walk in the alley, because mother wanted to avoid a particularly nosy neighbor who came knocking on our door. Mother simply did not want to discuss Daddy's problems with that woman. His absence while he was treated in a sanitarium lasted a week or two, and was distressing to us children. Our strong, silent father was unexpectedly revealed to be mortal, susceptible to illness which we didn't begin to understand.

So my father was a hard working man with limited income and long hours. He worked five and a half days per week, saved as much as he could, and was a faithful Catholic. Mass on Sundays wasn't even an issue, it was just a fact. As small children, we would kneel at his chair and recite our prayer every night ("Now I lay me down to sleep..."), and he himself knelt at his bedside every night.

Physical aches and pains included hearing loss, the single eye, and occasional back injuries. I remember seeing him crawl into the house after hurting his back while changing a tire. Nevertheless, he was stronger than his grown sons when he was 60, and maybe when he was 70. A lifetime of lifting furniture will do that, I guess.

Eminent Domain followed him around during the 1960's and 70's; the home of 30+ years was taken to provide playing fields for Vanderbilt University. The next home lost a segment of its back yard for a new Interstate Outer Loop and his place of business was taken for part of the Inner Loop. After losing the shop at age 70, he continued working, now solo, at home. On the day he died, he was warming himself at the gas fireplace, preparing to go down to the basement workshop, when he collapsed.

We were sent to Catholic school, grades one through twelve. At times, I felt deprived of neighborhood friendships, but the schools were good, and we had the benefit of lots of hours with the Religious Sisters of Mercy, including their various pointers, rulers, and paddles. I don't know what the practices were in public schools, but corporal punishment was lavished upon us whenever needed.

Money was never a topic of conversation in my household, but it seemed to me that we wouldn't be driving such an ancient car (a pre-war Plymouth) if we had any. (All through grade school, I just had to hope that none of my classmates would see that big, black antique waiting to pick up my sister and me.)

There were times when I resented my father for his discipline, but as I aged, I began to realize he was actually pretty supportive. As a Boy Scout my dad was always available to help with transportation. He was also present for any significant awards ceremonies, however boring he may have found them. Later, when I was 14 and had a paper route, he was always willing to drive me (at 4:30 A.M.!) when the weather was really terrible because of snow or heavy rains. He never had one word of complaint, which I noted even though I don't remember ever thanking him.

In high school, I "went out" for football as a Sophomore, landing on the "B" Team. Always slow-footed, there was never any chance of starring, but I was a varsity Guard in my Junior and Senior years. My parents came up with the money to let me buy letter jackets, with minimal resistance. I don't think I ever thanked them for that, either.

As noted earlier, Daddy was not inclined to pay for our schooling beyond high school. So college was a strictly pay-as-you go proposition for me and my brothers. Gordon, the oldest, worked a year after high school to save enough to get started, then worked at the University to pay his way. Jimmy, the next son, worked nights in a hospital pharmacy, where he could study when not actively engaged. The proximity of Vanderbilt University - within walking distance of our home - was a big help, but tuition had to be covered by our own savings. I made it through the first semester thanks to my accumulated wealth from a paper route (one year), work as a soda jerk at $0.50 per hour (two years), and then as a supermarket sack boy for $0.65 per hour (two years by the end of my first semester at Vanderbilt). Tuition was less than $1,000 per year, but I was short of money before the beginning of the second semester. I applied for a loan, but Vanderbilt evaluated our family income, as well as my academic record, and awarded me a grant for the second semester.

The grant was renewable, so long as I maintained a B Average, but participation in the NROTC program as an unpaid volunteer inspired me to apply to the Naval Academy - where tuition would be non-existent. My application required appointment by Congressman or Senator. As it turned out, my father was a grade school classmate (in 1904!) of a girl who later married our Congressional Representative. How could he refuse me a nomination?

Upon entering the academy, each midshipman needed to deposit $900 into an account which would be used to outfit him with books and uniforms. I still had $300 in savings, but my parents had to provide the rest. This they did without objection. Later, upon my graduation and commissioning, I repaid their contribution with monthly checks, and continued to send those checks even after I had repaid $600.

While at the naval academy, two months of each summer were dedicated to a sort of internship in some aspect of the navy. The third month was ours to do as we wished. In my case, that meant to return home. Part of that time was used helping my father. He especially enjoyed bringing me along to help deliver furniture. He took obvious pleasure introducing me as his son on leave from "Annanapolis". I must admit I enjoyed it, also - I was proud of his pride.

I've taken a long way to get there, but here's the point: I believe I've been shaped to a remarkable degree by this man I never talked to. In many ways, I never matched him. His physical strength, for instance, was greater than mine. His religious dedication was superior to mine, too. When he died in 1976, his bank account was found to contain $40,000 - much more than anyone expected - the equivalent of $215,000 today - and that memory has always nudged me to save, rather than spend foolishly.

Most importantly, I always remembered his unstinting willingness to help me as a child, and his quiet pride in my accomplishments. In particular, I tried always to be generous with my time and support to the activities of my children. I also am proud of their accomplishments. Financially, I wanted to help each of my children to pay for college, but expected them to take out loans as well. (One, more prideful than most, insisted on doing it all herself.) I'm especially proud of the fact that they each paid off their student loans as a first priority.

I learned some big lessons from my father. His steadfast strength despite adversity has been a guide to me. His religious faith was passed down to me. I have tried to follow him in his quiet support to his children, and his steady hand in financial matters. Maybe I could have done better, but I'm still working on it.

1970: Author's father and grandson, ages 72 and 6 months

29

The Brilliant Hayseed

I met Ronnie Gene Carter on the second ship of my navy life, when I reported for duty as Chief Engineer. Lieutenant (Junior Grade) Carter was the Main Propulsion Assistant – MPA – which means he was to be my primary assistant, serving as the Division Officer for both M and B divisions. "M" Division contained the Machinist's Mates, who worked on and operated the steam powered engines driving the ship. "B" Division contained the Boiler Technicians, who were responsible for the ship's four boilers. Both those realms were hot, dirty, and vital. Without steam, or without engines, USS *Floyd B. Parks* was just a 25 year old hole in the water, rusting at every opportunity, dripping oil and water (both salt and fresh) into the bilges. With any luck, the ship was attached to a pier whenever there was no steam. Underway or not, she was an old ship, and there was a spot in the hull, in the bilges directly beneath the after engine room control panel, where a pencil-sized jet of water could be seen when a temporary wooden plug was removed. Ronnie Gene relished this world, and the people who lived there.

The Machinist's Mates and Boiler Technicians (MM's and BT's in navy-speak), as well as others in the Engineering Department, were collectively known as "Snipes". These men were seldom seen by the Lords of Open Air who commanded the Weapons and Operations Departments. Snipe work was below decks, where noise, heat, and oil were omnipresent. Dungarees became ragged and oil stained, unlike the pristine starched dungarees of the operations and weapons specialists in the loftier reaches of the ship. Ronnie Gene was a Snipe. He was seriously overweight, with a belly overlapping his belt. His khaki uniform, while not actually ragged, was well worn, and did tend to collect oil stains.

Ronnie Gene was like his people in other ways. Frequently sounding like an uneducated hillbilly, he was prone to speech later parodied as Jimmy Carter English ("M R Farmers. M R Not! O S A R – C M Empty Pockets?!") "Them engines really need work", he might say. Or "We ain't gonna be ready for that inspection." This dialect went over well enough among the Snipes, but as an officer of the line, he also had to function among the wardroom officers in other roles. Like other Ensigns and JG's, he stood watches as a Junior Officer of the Deck. Duties included watching for other shipping, monitoring ship's position, and keeping station on accompanying heavies (Carriers and large Auxiliaries steaming in formation, for which the Destroyers provided protection). These duties involved reporting contacts and status to the Officer of the Deck, and sometimes to the Captain. Ronnie was able to speak with greater precision than his daily Patois, but in moments of pressure, out it would come: "Sir, that there contact just turned in our direction!" (Preferred: "Sir, target bearing 035 degrees relative is coming left – CPA close aboard to starboard")

Always a little tight and nervous among his peers, Ronnie was unusually susceptible to goading – especially from the Weapons Officer, who seemed to develop an unreasonable dislike for him. He commented frequently on Ronnie's size, occasionally on his diction, and always when he found an opportunity to criticize his watchstanding.

Nevertheless, I found myself developing a growing respect for Ronnie Gene. His people, under his management, kept the old engineering plant running better than many of our sister ships. The enlisted watchstanding, under his guidance, was sufficiently flexible and effective to operate as we needed to do – 24 hours a day, week after week. Initially, as we worked up for a West Pac deployment, I manned Main Control (the primary propulsion control position, located in the Forward Engine Room) at Battle Stations and all other special evolutions, while Mr. Carter went to the After Engine Room. But as I realized his competence at recognizing and dealing with actual or simulated engineering casualties, I was able to accept the Captain's invitation to become the Officer of the Deck for Sea Detail. This "dream job" enabled me to be on the bridge for all entries and departures from ports. I was the one to order "Take in All Lines", and to back away from the pier. Like most navy men, I enjoyed sea air and salt spray – both in short supply in the engineering spaces.

As time went on, despite the fact that Ronnie was unable to develop much camaraderie among the wardroom officers, I personally enjoyed his company. I also discovered that he was a lot smarter than we had thought. By the time the West Pac deployment ended, Ronnie Gene was a qualified Officer of the Deck, and had earned sufficient confidence from the Captain to take over as interim Engineering Officer when I left early to go to Vietnam.

Frequently a shipmate is a one-time friend – no matter how close you were on a ship. But fate allowed me to follow Ronnie Gene for a number of years. After I returned from Vietnam, I attended the Naval Postgraduate School, in Monterey California. There I found Ronnie Gene studying in the Operations Analysis curriculum – where one learns to apply mathematical approaches to solving real world problems. This was well beyond my academic skills (I studied Computer Systems Management), but Ronnie Gene excelled. We also visited socially. His wife, Julie, became a very good friend to my wife. In short, he became an admired friend.

Ronnie Gene's later assignments took him into the world of the Aegis weapon system – shipboard rocket-propelled anti-aircraft missile systems, incorporating a "phased array" radar system for both target detection and missile control. In short, this uneducated hillbilly had become a Rocket Scientist - first for the Navy, then for the developer, RCA. He never completely tamed his Missouri dialect nor his physique, but he ultimately was recognized and valued for his abilities.

EPILOGUES

30

Barbecue

I've eaten some strange things in my life, although they didn't always seem so strange to me. Example: peanut butter sandwiches with pickles - sweet pickles, or bread and butter pickles - never dill. Once, in Vietnam, I sampled pig gums, with teeth still attached. I did draw the line at fruit bat, however, and I declined to participate in consuming a soup containing a fresh-killed rabbit, despite several food-less days during survival training. I've sampled octopus and squid, though I reject raw fish.

In the vegetable realm, essentially nothing escapes - although some tastes received only short sample times. Once, when five or six years of age, I insisted on sampling a tiny green pepper. My parents warned me that it was hot, but my logic told me that such a cute little morsel couldn't be too bad, and besides my father liked it. The result: one seared tongue, from the tiniest morsel, which required a glass of soothing milk.

After that experience, it seems a little surprising (to me) that I soon learned to love barbecue, but once I tasted that first dish I was hooked for life.

I was not alone in my fascination with barbecue. My entire family, including my mother, participated. Furthermore, over time, I realized that a family reunion could not be successful without a heaping container of pulled pork barbecue.

For a few years, I didn't realize it had to be pulled pork. A periodic attempt around the family stove to produce the best BBQ sauce usually led to a paint job for chicken or beef. My father was generally considered the best maker of sauce. Ingredients always began with tomato catsup and vinegar. Usually some powdered mustard, lots of salt, and spoonfuls of both red and black pepper were added. Our magic cauldron was not especially huge - a two-quart sauce pan (well named!) was employed. I found this sauce exceedingly tasty - on bread. On meat, whether grilled or oven-baked, it had little effect. (Barbecue flavoring was wasted on beef or chicken, I soon concluded. When it comes to barbecue, I became and remain a lifetime pork devotee.)

Today, the wisdom of Wikipedia is available: "Barbecue or barbeque ... is both a cooking method and an apparatus/ machine. Barbecuing is done slowly over low, indirect heat and the food is flavored by the smoking process..."

Well, that secret of barbecue came as a slow realization to my family. There was an annual "fair" held each fall by the Knights of Columbus to raise funds for charitable works. One such charity was the Catholic boy's high school - the only one in Nashville. Probably some of the funds also went to the three girls' high schools. In any event, the fair was held on the grounds of Father Ryan High School, eventually the alma mater of all three boys in my family. The men of K of C set up a roasting a full day in advance of the festival, and cooked whole pigs over a long, slow, smoky fire. The Holy Grail was revealed! It turned out the barbecue sauce (similar to that developed in the Heins family kitchen) was a decoration, not the source of the flavor. Those pigs were continuously slathered with a vinegar and pepper concoction during an overnight roasting. I never had the opportunity to help with that vigil, but I assume a great deal of beer was also needed for the process.

As good as the fair's barbecue was, it only came around once a year. Something more reliable was needed, and soon enough, it was located in the shadow of another fair - the Tennessee State Fairgrounds. That venue was located about three miles east and south from our house (it seemed a lot farther then!). In the surrounding neighborhood were a number of low rent businesses - honkytonks, cafes, liquor stores, pawn shops and the like. One of those cafes did a thriving business in take-out barbecue. The Courseys had a full-time roasting pit, and prepared their barbecue in the time-approved manner. My father worked five and a half days per week at his own upholstery shop. He would occasionally stop there on a Saturday after leaving his shop and before coming home - bringing home the bacon, you might say. That barbecue became a family favorite, and established a standard of excellence for future reference.

Alas, all good things come to an end, and Coursey's eventually closed. Our family became BBQ refugees. Then, after wandering in the BBQ desert for several years, we discovered Browns' Barbecue. It took just one serving of that spicy specialty to convert me. As time went by, Browns' became the most important part of every family reunion and every summertime picnic. After a couple of years, Browns' was the gold standard of barbecue for the Heins family. Even youngsters being raised in such remote and un-southern corners as Michigan, Ohio, and Maryland soon swore allegiance to Browns', after being exposed to it in Nashville.

By the time I was old enough to drive and borrow the family car, the true faith spread among teen-age friends. An evening of beer and softball was perfected with a take-out order of Browns' barbecue. But it was always take-out, as there was no curb-side dining, and we could not dine in, as both the Brown family and their diners were "colored" and the year was 1957. Accordingly, my friends and I could only stand and wait for our order to be handed over. (Since this was the way it had "always" been, and always would be, it never occurred to us that it was strange. Even the Greensboro lunch counter sit-ins were still three years in the future.)

One defining moment, established in my late teen years, involved the Brown's barbecue sauce. I came home late at night, with some precious pulled pork left, along with a Styrofoam cup of sauce. I felt the meat needed refrigeration, but not the sauce, so I left it on the dining room table. The next morning, it had burned through the cup, leaking onto the table. That moment lived on in my family as testimony to the potential strength of the sauce.

In 1959 I left home to attend the Naval Academy - 24 hours away, by Greyhound Bus, and so began The Long Hunger for barbecue. Not once during my four years by the Severn did the powers-that-be serve barbecue in the mess hall. The hunger was broken over the next 20-odd years by visits home and the occasional family reunion, usually but not always in Nashville. In foreign locations, nephews (although not nieces) would scour the countryside to find the nearest equivalent to Browns', but the gold standard was never displaced. (More recently, my son, while stationed in Hawaii, mounted a strong challenge using Wikipedia, his slow cooker, and "liquid smoke" - but that's a little too far to go for BBQ.)

Years passed, and I came to Newport, home of the Naval War College. Despite the small town proximity of life, it took me several more years to discover... Becky's Barbecue! That same familiar tang, with its sweet touch of smoke, of Tennessee barbecue! The sauce here doesn't burn through the (now plastic) cup, but the taste is passably similar, despite the lack of sulfuric acid.

The new challenge, if I may call it that, is to avoid trivialization. Now that high quality BBQ is available any time from a reputable store specializing in it, it's almost too easy. No sketchy neighborhoods, no long treks, no budget-busting costs.

It's so easy it may not be worthy of its exalted culinary status. Ah, well - that's probably the way super-rich people feel - the challenge and the tang are gone from life - but who would willingly change it?!

31

Radio Days

Keeping the Kids Entertained

I WAS BORN AND raised in Nashville, Tennessee. Later it was known as the home of country music, but in my earliest days, that would have been a pointless boast - "Who cares?" would have been the obvious response to the claim. Country music was being recorded and performed there, but it was an obscure field, of interest primarily to uneducated farmers across the south who listened to *The Grand Ole Opry* over WSM - one of two 50,000 watt "clear channel" radio stations in the city.

For our family, radio offered some opportunities for very inexpensive (as in free) entertainment. Some radio shows were created in large studios, where there was room for an audience of maybe two hundred guests. One in particular was a "breakfast show", not that different from today's TV offerings like Good *Morning America* or *Today.* There was always an orchestra, a Master of Ceremonies, and radio personalities. I remember once being there for an appearance by Lester Flatt and Earl Scruggs - a bluegrass duo who continued to flail the banjo and mandolin for many years to come. Interestingly, their radio personas were hick farmers in overalls, but they were dressed in business suits for the radio performance - despite having an audience.

My mother was remarkably resourceful in keeping us kids occupied and entertained. In addition to our rare visits to the

Morning Show, we visited the Noontime Neighbors show more frequently. It was pretty similar to the *Breakfast Show*, except for the time. They did have a ten- or twelve-inch bell which was sounded at the beginning of the show (12:00 o'clock sharp), with the call "It's noontime, neighbors!" - like a prosperous farm calling in the hands for a noon meal.

Other shows were created in much smaller studios, but there still was a small room for performer, announcer, and visitors, separated by a glass window from the engineer. Once we visited the broadcast of the Eddy Arnold show. He later became a real super star, but in the late 1940's his success was just beginning, and he still did a live 15 minute show on a daily basis.

Another studio visit allowed me to watch the "re-creation" of a baseball game. A local station, WKDA, broadcast every game of the Nashville Vols - a Double A team in the Southern Association. I was a huge fan, attending all the games I could wheedle out of my father. He must have been a fan, as well, as he took me to see this broadcast re-creation. The announcer, Larry Munson, was inside one of those glassed-in studios, with the engineer. They had a long-playing vinyl record of crowd noise. At low volumes, it was simply the murmuring of a large crowd. When cheering was indicated, he just turned up the volume. He had a slab of marble, and a small wooden "bat" wrapped in a towel. There was a ticker-tape narrative giving him the barest facts, and he would elaborate, giving the impression he was watching the game in person. The primitive sound effects were very effective!

Radio shows were not our only entertainment, of course. Mother would frequently take us (mostly me and my younger sister, in my memory) downtown to the library. Our trips were by city bus - an adventure in itself. At the library, I always made sure to get my limit of books. I especially liked books like *Daniel Boone, Boy Explorer*, and *Thomas Edison, Boy Inventor*. Yes, there was a certain pattern there. Another favorite at the library was a marionette show, professional in quality, with engaging stories and fabulous puppets.

Time at home, also, frequently included radio. Some programs were institutions in their own way. On Saturday nights the family would gather to hear Gangbusters, introduced by the sound of sirens and machine gun fire - very exciting - and a precursor to today's TV police procedurals. Sunday evenings always included carefully monitored episodes of the Jack Benny show and the Phil Harris show. Each of these was a mixture of comedy, music, and loose story lines - really like nothing currently on the air, but very much like some shows in the earlier years of television. Other shows in the evening, still with me somewhere, were *Suspense*, *The Shadow*, and a Sci-Fi anthology called *Dimension X*. There were news and public service programs as well, but naturally, I paid them less attention than did my parents.

Weekday afternoons, a number of shows were aimed at us kids, and we did pay attention. Typical was *Superman* - a fifteen minute show, as I remember, which came on about 4:15. We neighborhood kids would stop our play and come into one of our houses to hear the latest adventures of the Man of Steel. Other shows existed, and were less captivating, but we would listen in wintertime or days of bad weather. These included some shows which live on in my memory - *Sky King*, the *Lone Ranger,* the *Green Hornet*, and *Sergeant Preston of the Yukon*. Scenes from those shows remain more vividly painted in my memory than anything I ever saw on television.

When I was eleven or twelve, television was just arriving in Nashville. My parents went shopping for (or so I thought) a TV, but wound up buying a console model AM/FM radio with phonograph. It was a disappointment to me, but my parents felt the cost and current state of development simply did not warrant purchase of a TV. Disappointed or not, those Sunday evening radio shows were enhanced. Furthermore, I developed a conception of what TV would be like - I imagined peeking into the dial, into a cavity where all the characters were performing on a tiny stage - each was about six inches tall. When we finally did get a TV (just in time for the coronation of Queen Elizabeth II) The fuzzy black and white portrayal could have been quite a disappointment - but it wasn't. We would even watch the test patterns when the broadcast day had not yet begun.

Radio never died, but the radio dramas did. I listened to baseball right through my teen years; the local team never was telecast, but that didn't seem to matter. Music became very important to me as a teenager - Rock 'n Roll originated, as I remember it, about 1952 - when I was 12. And of course, country music was increasingly popular - no longer just for farmers. Gradually, television supplanted radio, and became the new king of the hill - for a time. Now, of course, computers seem to be creeping up that same old hill.

32

Tom Mix and Me

Anyone my age (I'm fully adult, to put it euphemistically) remembers radio dramas. They were pretty similar to the TV dramas today, but shorter. Some ran just 15 minutes, but those were on every day, so we weren't shortchanged. Most of the action was presented through dialog, although sometimes there was a narrator.

One sub-genre was the radio cowboy. One notable horse opera was *Gunsmoke*, starring William Conrad as Matt Dillon. (1952 - 1961). That particular show was perhaps the last of its kind, but it was one of the few which transitioned into television where it ran for some 20 years (1955 - 1975) - despite the disappointment we transitional fans felt when we lost the resonant baritone of Mr. Conrad. Over the years - in fact quite quickly - we learned to accept the laconic speech of the new Matt Dillon, James Arness. That TV series continues in reruns to this day.

But long before Conrad or Arness, there were many other riders of the purple sage. In early childhood, there were Gene Autry and Roy Rogers - both "singing cowboys", but primarily appearing in movies, not on radio. At age 6 or so, I had a Roy Rogers outfit, consisting of vest, six guns, and chaps. There may have been a hat, too, but I have no memory of it. Later on, after TV came to our land, there was Hopalong Cassidy - the good guy in black. And who could forget the Lone Ranger?

An early radio favorite of mine was Tom Mix; he really existed, born in 1880, a horseman from youth, though never a workaday cowboy. He made more than 160 cowboy films, mostly silent movies. My father (born in 1898) was a fan in his own boyhood. His radio series ran from the 1930's to the early 1950's, called "The Tom Mix Ralston Straight Shooters". I didn't know it then, but Tom Mix himself never appeared on that radio series (as a matter of fact, he died in 1940). I listened faithfully during the mid-40's, and responded to more than one special radio offer from Ralston-Purina, the sponsor. Despite having never owned a cowboy boot, or a horse to ride, I felt it important that I should own a pair of glow-in-the-dark spurs. I dutifully filled out the form on the sponsor's cereal box (possibly Ralston Chex), cadged a half dollar from my mother, and mailed it in. In due course, those spurs came, carefully packaged and ready to ride. After exposure to sunshine or a bright light, they could be taken into a closet and admired. Emitting a nice greenish glow, they clearly exhibited their usefulness in keeping track of your friends while setting up a nighttime ambush for outlaws.

Not too long after obtaining the spurs, a new product, better than ever before, was offered. This time, it was a glow-in-the dark arrowhead, with built-in compass and magnifying glass. Yes, it required another fifty cent investment, and another agonizing wait for Battle Creek Michigan to deliver, but the incomparable allure and utility was clear. So, once more I filled out the order form, obtained a fifty-cent piece, and sent my order out in a three-cent first class envelope.

Thus began a wait I could never have anticipated. Weeks went by, but I had been warned that delivery could take up to six weeks. When weeks turned to months, my interest flagged, then faded. A tiny, bitter hole developed in my psyche. Eventually, I stopped listening to the Tom Mix radio show, and that bitter hole hardened into a grudge.

I moved on, as children do, and lived my life. School became more central, soaking up many hours each week. Radio shows became less central, then were eclipsed by television. Bicycles and friends became important.

Although "they" say "'Tis better to give than receive", I never really subscribed to that. I enjoyed Christmas presents, birthday presents, Halloween candy, and penny purchases of treats at neighborhood stores. I did not again send off for treasures endorsed by a breakfast cereal, as I remembered my grudge.

One memorable birthday present was the baseball I received at age ten or so. A simple thing, but it was special to me. My birthday is in March, and I was a (Nashville Vols) baseball fan, so there was just something wonderful about having a brand new baseball at the beginning of Spring.

When I was twelve, my primary Christmas present was a Cocker Spaniel puppy. I had no idea, though I learned afterward that she (Pepper) had cried for attention while being confined to the basement during Christmas Eve. My older siblings went out of their way to distract me and make noise, but I never noticed a thing, so it was a complete surprise when I went, early Christmas morning, to the fireplace hearth where Santa's presents invariably materialized. Joy, indeed!

Many years later, my adult children surprised me at my daughter's apartment in Oakland, when my wife and I were passing through (on my birthday) from a cruise around the Hawaiian Islands. That was special! It was a gift, because it took place on my birthday, and was completely unexpected. Beth lived in Oakland (we were stopping over for a day or two). Mary lived in Orlando, John in Baltimore. The sudden appearance of all these children was a very special birthday present.

In 2015, there were two major gifts: One was a flight in an ultralight aircraft in Hawaii. John now lived there, thanks to the Air Force. He had married a year earlier, so my new daughter Kate was also there. Beth and Mary joined with John to fund the ride; they must have shared a very hefty fee, and it was a superb experience - much better than a helicopter or glider ride could have been. The other very special 2015 gift was a "Worlds' Best Grandpa" mug, which announced the impending arrival of my first grand child - special indeed!

Throughout all these years, and whenever the topic of "gifts" arose, I must have told the story about my grudge - the one that got away, the Tom Mix glow-in-the-dark arrowhead. I wasn't really aware of the repetition; it was just one of those family stories told by old folks everywhere. Eventually, the suffering children probably could have recited the story verbatim, after the first three words were out. Well, in 2016, son John shocked and delighted me with a birthday gift for the ages.

A few days before my 76th birthday a package arrived, "forwarding" a RALSTON STRAIGHT SHOOTERS envelope addressed to Little Raymond Heins, at 2513 Capers Avenue, Nashville Tennessee. (Even the street no longer exists, let alone the house.) In the envelope was my long-awaited Tom Mix glow-in-the-dark arrowhead! I had abandoned any hope that such a thing even existed, 69 years after I ordered it. But John was a hero, found it on e-Bay, and fulfilled my long-frustrated dream. It was exactly as advertised, with magnifying glass, compass, and even "Tom Mix" embossed in the plastic.

As birthday gifts go, this one will be hard to beat; Thank You Hero John!

33

GO HOME AGAIN, AGAIN?

In what now seems a distant time, I mused the topic "Can You Go Home Again?" The focus was on Christmas, and upon reflection I opined that you can, indeed, go home again:

Quite clearly, the truth is, home is where your heart is. There can be a different "Home" for every member of the family, but I will always believe that each of us can, indeed, go "Home" again!

This year, under greatly modified conditions, I felt it was time to consider the question again.

In June 2021, my spouse and I sold our Middletown home of nearly 50 years, and moved into St Clare, an Assisted Living residence in Newport. We traded in our relatively spacious Raised Ranch for a two bedroom apartment, with one bath and essentially no kitchen, and a completely different life style. Now, someone else was responsible for maintenance, heating/cooling, cleaning. No longer did we need to provide, prepare, or shop for meals. Instead, we have a congregate dining room and 20 or so fellow diners. None of our fellow residents were relatives, nor even acquaintances. Most of them were elderly and infirm, with an average age of 90. As an 81 year old, I was an elder in training, and my wife is even younger than I am. Nevertheless, we soon found our new housemates to be charming, welcoming, and wise. Our residential facility does accommodate more limited residents, but the levels of care are segregated.

In December, 2021, we flew to Texas, the temporary residence (thanks to the USAF) of our child with the largest house (and also a wife and two grand children). All previous family Christmas seasons had been in our residence, beginning in 1970 when there was just one child. In 50 years, a number of customs had developed, but the over-riding concept was that we were Home for Christmas.

In 2021, many things were different. We needed multiple hours in airliners to arrive, for one. Another shock was the last-minute notification that our first born, who lives in a Group Home in Barrington, could not travel because all five residents had tested positive for Covid illness. Michael could not come with us to Texas.

Our son and his family (John, Kate, Clara, Henry) offered wonderful hospitality. Our two daughters, both residents of Portland, Oregon, were their charming selves, doing all they could to help make it a wonderful Christmas experience. Still, it was Texas! Temperatures were of the shirt sleeve variety. Scorpions were more likely to be encountered than icicles and snow cover.

The visit itself was grand, and I fault no one for any lack of welcome or cooperation. Christmas Eve Dinner included a gathering of both families (John's and Kate's) as well as the grand children – always welcome and perfectly behaved, in my (possibly biased) opinion. Wine and fine dining were beautifully served. Christmas Eve included a mass, mercifully early and not terribly lengthy, but somewhat over-heated. Santa had no problem finding the house, and he did a fine job of choosing. Christmas Day included the ceremonious opening of presents, and a 20 minute drive to the residence of Kate's parents, where more wine and more food were the order of the day.

So what kept me from enjoying all the festivities just as I would have at home? In retrospect, I was developing a scratchy throat and an icky cold. By the time we headed back toward

New England, two days after Christmas, I was moving from icky cold toward pneumonia.

On the planes traveling home, I slept all the way. We had paid a premium fare, to guarantee spacious seats and first class meals, but I found it impossible to touch the food. Perhaps I took advantage of the roomy seat, but I was not aware of it.

There was a plane change in North Carolina, and we managed to lose my wife's Walker – it took a little trip of its own, and arrived at our house two days later, But that was just the beginning of tribulations.

At the Providence airport My wife and I entered the Twilight Zone. I could not find the parking ticket for our car. Despite a foggy brain and chronic back pain, I went looking for the car, and actually found it. Unfortunately, with no ticket (and no cell phone – I had unintentionally left it behind), I could not extricate my auto from the parking garage. Walking back to the terminal, I waited a half hour to meet an airport policeman who very politely drove me around until we found the car. He was able to get me out of the now closed parking garage – a kindness I still do not under-estimate. But with my foggy brain increasingly in control, I left behind in the patrol car my handbag containing a computer and, once again, my cell phone.

Nevertheless, we headed out, after midnight (in Twilight Zone time), to drive home. There my brain fog reached stage three, and we took a fifty mile tour of a ghostly, foggy, nearly vacated Rhode Island. Evidently I had not mastered its highway system despite 50 years' residence. We did have a GPS in the car, but one with a very poorly developed interface. (My cell phone would have solved that problem, but Rod Serling had confiscated it.) Guess what: there are many Spring Streets in Rhode Island! I remained lost in RI, even returning to the airport at one point, until finally it did occur to me that I could ask the GPS to find "St Joseph Newport", so we were able to find the Newport Bridge.

Tortures were still not over, however. After arriving in our apartment, the on-duty nursing staff determined that my blood oxygenation was alarming, and I was transported to the Newport Hospital ER for a multi-hour visit. The hospital diagnosed my pneumonia, and determined that I needed both antibiotics and anti-viral medication. Surprisingly, they could not find a positive Covid test. In fact, I was tested nine times in ten days, and never did come up with a positive result.

Despite my lack of Covid, the problems were not ended. The spasmodic coughing spells threatened to strangle me. Furthermore, paranoia and hallucinations developed. I was convinced that hostile Canadian Railway forces were injecting coughing agents into my environment. At one point, I realized that my wife and I needed to escape from those hostile forces, and that I had keys to a car, in which to escape. She talked me out of that, fortunately. Nevertheless, each time I went to sleep, I awoke with gasping, spasmodic coughing, hardly able to breathe. Eventually, I realized that if I sat up, the coughing was less likely to engulf me.

Finally, some combination of position and meds seemed to reduce the coughing. The staff nurse came into to our apartment check on me. Having just awakened, I explained to this lady about the agents who persisted in trying to kill me with coughing gas. As I spoke, I saw an expression on her face, which I now find rather funny; she was looking at me as if I had three heads; probably she was calculating where she could obtain reinforcements if I decided to make a break for it! I just stopped talking, a few minutes too late.

Eventually I obtained effective treatments for the coughing, and began to recover – somewhat. Even so, that nurse ruled that I couldn't return to the congregate dining room for what turned out to be two weeks. "You wouldn't want the other diners to get what you have", she said.

And so it happened that my wife and I did hard time –

solitary confinement – until we'd been home from Texas for a month. As both my health and sanity began to return, I kept thinking about the differences between this Christmas and the fifty before it. Finally, I did reach a conclusion: Maybe you really cannot "Go home again" if your health and mind desert you.

About the Author

Raymond Heins was inspired to write his memoir, *Ripples*, after attending the "Circle of Scholars" writing course taught at Newport's Salve Regina University by professor and author Jack Galvin. The program included a series of writing assignments under the "Write Life" umbrella. Essays focused on self-discovery and prompted students to delve deep into their history, psyche and personal experiences. Ten years later, *Ripples* was ready to publish.

Many of the episodes in Heins' memoir sprang from his diverse military career. After graduating from the Naval Academy in 1963, Heins took up Navy's promise to "Travel the World" and steamed to the Mediterranean, the Middle East, and Northern Europe, followed by tours in the Western Pacific and a on a destroyer in Vietnam waters. Heins next spent a year "in-country" as the American advisor on a Vietnamese gunboat.

Following Viet Nam, Heins attended Naval Postgraduate School where, in 1970, he earned an MS in Computer Systems Management. Later, serving on a team at the Defense communications Agency in Reston, Virginia, Heins helped develop one of the early inter-computer networks for senior military commanders around the world. As he says with a wink, "this was when Al Gore and I invented the Internet."

After a five year stint at the Naval War College, and retirement from military life in 1983, Heins joined the industrial world and eventually the Computer Sciences Corporation, working on war gaming computer systems until 1994. That was followed by yet another adventure when Heins entered the world of Rhode Island real estate for thirteen years, including eleven as an appraiser.

Ray and wife Virginia Blake (of more than fifty years) have four children and, when health issues permit, participate in civic and church activities, remaining in Rhode Island at a Newport Assisted Living home. As Heins' diverse passions spurred him to reinvent himself yet again,

Ripples is his first adventure into Ray's own "Write Life".

Made in the USA
Monee, IL
13 September 2026

62976633R00098